Articulating Publicness in Infrastructure: The History of Municipal Streets, Water and Sanitation in Sweden

Articulating Publicness in Infrastructure: The History of Municipal Streets, Water and Sanitation in Sweden

Pär Blomkvist

Published by   **IWA Publishing**
Unit 104–105, Export Building
1 Clove Crescent
London E14 2BA, UK
Telephone: +44 (0)20 7654 5500
Fax: +44 (0)20 7654 5555
Email: publications@iwap.co.uk
Web: www.iwapublishing.com

First published 2023
© 2023 IWA Publishing

British Library Cataloguing in Publication Data
A CIP catalogue record for this book is available from the British Library

ISBN: 9781789063974 (paperback)
ISBN: 9781789063981 (eBook)
ISBN: 9781789063998 (ePub)

This eBook was made Open Access in August 2023.

Doi: 10.2166/9781789063981

Contents

Chapter 10
Comparing publicness in municipal infrastructure

About the author

Pär Blomkvist is Associate Professor in Industrial Economics at Mälardalen University (MDU), Sweden. He holds a PhD in history and his research focus is on infrastructural systems and innovation theory. Blomkvist has published several books and articles on infrastructure management in road traffic, and water and sanitation systems, from both a historical and a contemporary perspective.

doi: 10.2166/9781789063981_xi

Foreword

This is a book about the history of municipal infrastructure in Sweden. Since I started the project, the gravity of the upcoming climate crisis and its effects on infrastructural systems and society, has become even more evident. The speed of climate change is surprising and our future seems precarious. While finalizing the book in June 2023, Sweden, and many other countries in Europe and all over the world, face yet another extremely warm summer with water shortages, crises in agriculture and burning forests, not to mention deforestation, rising sea temperatures, melting polar ice, wildlife extinction and Russian ecocide in the Ukraine. Moreover, erratic and heavy rainfall, flooding and the climate running amuck are not distant prognoses any longer, but a harsh reality. Infrastructure development is pivotal in adapting society to the challenges ahead. To mitigate climate crisis effects, at least to some extent, municipal infrastructure, streets, water and sanitation systems need to be refurbished and strengthened. My hope is that, in some small way, insights from history can help in this endeavor.

The book is published by International Water Association Publishing (IWAP). It is based on a research report within the project InfraMaint funded by MISTRA (The Foundation for Environmental Strategic Research), hosted by RISE (Research Institutes of Sweden). The research report is posted in a digital format, as a compilation of knowledge and a reservoir for further publications, under the title *Research report and excerpts on the history of municipal streets, water and sanitation in Sweden* on this website: https://mistrainframaint.se/publikationer/. In the following, I refer to the research report as Blomkvist (2023a).

The reference list includes only the most important literature used. Primary sources and web sites are not included. The list is condensed and adapted to an international audience. The research report, on the other hand, presents lots of raw data in the form of excerpts and quotes from literature and primary sources and contains all relevant information in the form of footnotes and a full reference list. The report has a similar structure to the book, so it should be quite easy to find the appropriate reference, side numbers, etc.

Acknowledgments

The book is part of the research project InfraMaint founded by Mistra (The Swedish foundation for strategic environmental research) and hosted by RISE (Research Institutes of Sweden). Thanks to Magnus Arnell, Lars Thell Marklund and Åsa Flydén at RISE for all the support I could have wished for. I also want to acknowledge the help from participants in InfraMaint who have commented on my work at seminars and meetings.

The book proposal to IWAP was peer reviewed by four anonymous reviewers and I am grateful for their comments and for the generous help from Mark Hammond, Books Commissioning Editor and Katharine Allenby, Production Editor at IWA Publishing.

The original research report has been peer reviewed by many. I have had the pleasure to discuss my work and received a lot of help from the following experts in infrastructure and historical studies and most of them have at various stages of completion read and commented on my drafts: Uno Jakobsson, Erik Winnfors, Ove Pettersson, Hans Bäckman and Erik Karlsson, Svenskt Vatten; Jonas Hallström, Linköping University; Erik Lindberg, Uppsala University; Lars Nilsson and Håkan Forsell, Stockholm University; Jesper Larsson, Swedish University of Agricultural Sciences; Martin Emanuel and David Nilsson, Royal Institute of Technology. I thank Jonas Christensen, Peter Ridderstolpe, and Björn Eriksson who generously shared their knowledge on small-scale sanitation at a special seminar in Uppsala, April 2023. Moreover, Christensen has also shared his deep knowledge in the complex area of water and sanitation legislation. Thanks also to Fredrik Petersson, Stockholm University, who gave me a copy of Wilhelm Leijonanker's plan (1853) of the piped water system in Stockholm; Henrik Kant, the City Planning Administration, Gothenburg; and Ulf Thysell, the Municipal Water and Wastewater Company in Northwest Skåne (NSVA) who helped me to see outside a narrow Stockholm perspective. I am also grateful for the help from Albin Blomkvist in revising the footnotes and the reference list.

A special thanks goes to Mats Hallenberg, Stockholm University, who commented extensively and helped me to understand the concept of *publicness*, and again, to David Nilsson who has shared his deep knowledge on water and sanitation (WS) history and been a valued discussion partner. David also read the book in its final stages and gave valuable advice and constructive critique. All remaining mistakes and misunderstandings are my own.

Water and sanitation were relatively new topics for me before this project. My foremost inspiration in WS research comes from Jonas Hallström, who has written the most comprehensive Swedish historical case study of the introduction of piped WS, and from American and Swedish doyens in WS history, Martin Melosi and Jan-Olof Drangert.

Kärrtorp, Stockholm, June 2023
Pär Blomkvist
Associate Professor in Industrial Economics at Mälardalen University (MDU), par.blomkvist@mdu.se

Keywords: Policy and governance, Utility/network management, Water supply and treatment, Infrastructure history, Roads and streets, Water and sanitation, Large Technical Systems

Abstract

The first contribution of the book lies in the historical comparison of infrastructural systems that normally are dealt with separately. The synthesis has been achieved mainly by an extensive literature review of research from a wide range of various fields and by using prime sources to some extent. Earlier results have been reinterpreted and research areas that communicate rarely have been brought together. The comparative and long-term perspective allows the discovery of similarities and differences in the development of arrangements around streets, water and sanitation (WS). By using the analytical lens of *publicness*, the book challenges the common belief that these three areas have always been public concerns or obligations. An assumption based on the fact that presently they indeed are *public* infrastructural systems. The second contribution is the connection of the historical development of these three sectors with research in medical, social, cultural, economic, technical, and political history highlighting the most important contextual factors in society at large that has profoundly affected streets, water and sanitation. The book shows how their respective evolution into public infrastructural systems has been strongly influenced by the strong Swedish tradition of local independence, by urbanization, demography, and industrialization, the municipal reform of 1862, and specifically for WS, the divide between the private and the public spheres, the social issue (and fear of cholera) and new perceptions of health and sickness, the Sanitary movement, and the National Health Act of 1874. In the twentieth century, two additional contextual factors influenced the articulation of publicness: first, environmental concerns due to water pollution and second, circularity (reuse of resources) and sustainability. Finally, the book shows how the evolution in municipal streets, WS has left a historical *legacy* still affecting the way these infrasystems are managed today.

doi: 10.2166/9781789063981_0001

Chapter 1
Starting points

In this chapter I present the foundations and delimitations of the research project presented in the book, the most important concepts and methods used, as well as the general layout.

1.1 INVESTIGATING PUBLICNESS

Since the dawn of civilization, some sort of arrangements has been made to provide for roads (called streets in towns), drinking water and sanitation (WS). In one form or the other, people have always strived to cater to these essential resources and the field is huge with a very long history. My focus is on the provision of streets, water and sanitation in Sweden from roughly the beginning of the nineteenth century, which led to the creation of modern municipal infrastructural systems, also called *infrasystems* (Kaijser, 1994). However, I will account for parts of their respective pre-modern history, which still influence the systems today and discuss some international aspects. The history of municipal infrastructure in Sweden is intimately connected to European and American development.

I am writing a *comparative* history of municipal infrastructure and investigate how these fundamental service *arrangements* became publicly managed infrastructural *systems* in the end of the nineteenth century. We may believe that streets, water and sanitation have always been public concerns, but that is not the case. History reveals a process of transformation into the public realm closely connected to contextual factors affecting society at large. I call this process the *articulation of publicness*.

Today we normally think of *two* infrastructural systems, roads and the water and sanitation system, respectively. But in a historical investigation, this line of thought is misleading. In fact, we have *three* modes of service *arrangements* turning into *systems*. The gradual *systemization* of these three areas shows different paces, motivations, ownership forms and technical and institutional designs. It is important to take a step back and recognize the character of these arrangements and infrasystems and what type of recourse or service they provide. Thus I distinguish between the terms *arrangement* and *system* to highlight the evolutionary process of transformation from older modes of *off-grid* service *arrangements* into modern *on-grid* infrastructural *systems*. Although not used in earlier research, this distinction is fundamental to my argument. Historian Joel Tarr, for example, uses the term *system* both when talking about the old (off-grid) 'cesspool-privy vault *system*' and the modern (on-grid) 'water-carriage *system* of waste removal' (Tarr *et al.*, 1984). In this book, the term *system* refers only to service arrangements that have taken the form of modern infrastructure.

Another important distinction is that what we see is not a complete and total transformation from arrangements to infrasystems. In the road sector, we still have a local level of civic roads managed by the nearby property owners. Although I argue that these roads are relatively well aligned with the roads and street system, they are not entirely turned into a part of a modern infrasystem. Regarding water and sanitation, we do not even see this grade of alignment between the local level and the system. In water and sanitation the local level still includes off-grid and pre-modern service arrangements which are not particularly well aligned with the system.

The main question in the book is how service *arrangements* of streets, water and sanitation turned into public infrastructural *systems*, and how they gradually were *articulated* as public responsibilities. By *articulating publicness*, I mean a process whereby an area in society which earlier has been defined as being part of the private sphere, a task for the individual citizen or the household, is transformed into a task where public bodies such as the state or the municipality carry the responsibility. This does not automatically mean that a public organization must implement the tasks by establishing a special business organization. It is possible to have for example public gasworks run by commercial private companies, and as will be discussed later, it is also possible to have private individuals such as property owners and farmers performing the public tasks of road and street maintenance. In conclusion, the level of publicness is not a given based on intrinsic properties of the resource or service in question. Publicness is politically constructed and dependent on the actual historical context. It is *articulated* as such. In other words, using the Latin vocabulary discussed later, the focus is on whether streets, water and sanitation are *Res Publica* or *Res Privata* (public or private affairs) rather than if they are intrinsically public or private goods.

The book is about *municipal* infrastructure, but it is mainly centered around towns. The countryside gets less attention, except concerning road and street history, simply because the development toward modern infrasystems started in these densely populated places and most of the book is about the first formative decades of infrastructure evolution. Another more pragmatic reason is that towns have been much more discussed in earlier research. For the same reasons, the book is mostly focusing on the capital Stockholm and to some extent larger Swedish towns such as Gothenburg and Malmö, especially when it comes to water and sanitation, because these cities were forerunners in infrasystem development, and most towns followed their, and especially Stockholm's, example in the construction of piped WS. The terms *town* and *city* are used interchangeably.

Roads, which as mentioned, are called *streets* in towns, have been vital for society since time immemorial and have always been closely related to societal matters such as trade, warfare, and the national territory. Road and street history is a quite straightforward, but slow and sometimes erratic, movement toward a (relatively) unified national, and even transnational, infrastructural system.

The history of drinking water is not as straightforward though. This is partly because water is such, and I apologize for the pun, a fluid concept. It has many meanings and is used in many types of situations. When we talk about water and investigate its history in relation to humankind, we can for example mean the ocean (salt water), rivers, lakes, or ground water (fresh water) and focus on potable water provision, irrigation, sea transport, fishing, waterpower, industrial use, or pollution of water recipients. From a natural scientific perspective, we can investigate the so-called *global water cycle* and follow its journey from evaporation to its return to water bodies or ground water aquifers. In sanitation, as will be discussed below, water is used for the transport of excrement in piped sewers, for street cleaning, firefighting and of course for household hygiene, cooking and washing, and so on. In the following, I label the provision of drinking water and sewage as the *local water cycle*, that is, the extraction from water source via purification, distribution, use, wastewater treatment, and its return to recipients. Thus, I turn to water provision, of fresh drinking, so-called potable, water, and investigate the history leading up to the introduction of modern water systems in underground pipes at the end of the nineteenth century. When writing about water in the following, if nothing else is specified, I mean drinking water.

Fresh water is vital for humankind. We simply die if we don't get it and water quality is essential for health, at least we know that today. Fresh water has furthermore almost always been associated with positive and sometimes spiritual and magic qualities, and at present, water is often seen as a *human right* following a UN resolution from 2010 (GA res. 64/292). However, there are historical indications that drinking water was not always appreciated. Some researchers, for example, claim that the Roman elite saw water as the characteristic drink for the lower classes, slaves, women, and children. Free *men* of Rome drank wine and that the aversion to water was carried into the Middle Ages and the pre-modern era:

> … Drinking water – any water – was a sign of desperation, an admission of abject poverty, a last resort … In the seventeenth century Europeans generally disliked, distrusted, and despised drinking water. Only truly poor people, who had absolutely no choice, drank water … There is one thing Europeans agreed on: drinking water was bad – very bad – for your health. (Saltzman, 2012)

This argument can be found also in a Swedish context, and perhaps the evidence of a very high beer consumption in the seventeenth and eighteenth century could point to the same aversion. Everyone just knew that water could make you sick to the stomach. One example can be found on the island of Gotland on the Swedish east coast: in October 1876 when the island was hit by a typhoid fever epidemic. One city doctor realized that more children than elderly had fallen ill, especially among the poor, and concluded that the reason was probably because, according to the habit of the area, the adults drank light beer (svagdricka) instead of water. Beer was of course also a way to process and preserve grain and an important source of carbohydrates. The negative attitude toward drinking water changed somewhat in the nineteenth century when water from wells and springs came to be seen as healthy and invigorating.

Dealing with sanitation has of course also been a vital concern, especially in densely populated dwellings. However, the handling of excrement has mostly been seen as a strictly private concern and a necessary evil surrounded by all sorts of taboos (Black & Fawcett, 2008). The term *sanitation* has had a lot of meanings in the course of history, and I will discuss these later. It is worth noting that *sanitation* often is used when discussing all sorts of waste removal in historic times, which can be a bit confusing. In modern day language, referring to piped sewers, the terms *sanitary sewer* and just *sewer* are used for wastewater and feces and *storm water sewer* for the removal of excess rain (storm) water. To complicate matters even more, sewage and storm water are often transported in the same pipe, so-called *combined* sewers. But the confusing vocabulary and complex institutional framework is not only a historical phenomenon. As will be discussed, the *patchwork* in legislation and the many organizations are exceedingly topical in the current water and sanitation sector (Christensen, 2015).

It is evident that streets, water and sanitation, presently managed as public infrastructural systems, are fundamental for modern life. They are in fact so basic that we often take them for granted. Our roads and streets have turned into publicly managed paved corridors for automobiles, drinking water pours out of the tap, and the handling of excrement has been moved away from us by piped sewers. We don't think about these basic services, we don't have to work to maintain them in our daily lives, and the only time we notice the infrastructure is when it breaks down or malfunctions. According to a recent study of municipal streets and WS, this *invisibility* is attributed to a 'deep taken-for-grantedness' because infrastructures are prone to fade away from conscious awareness: 'In turn, a sudden absence of, or a dramatic change in, the flows render the underlaying infrastructures visible and the everyday functional aspects of infrastructure become apparent' (Alm *et al.*, 2021; Blomkvist & Kaiser, 1998). However, it must be noted that this picture is only true in some parts of the world. In many low-income countries, the services provided by modern infrastructural systems are still a hardship needing manual labor. Roads are not kept by public organizations; excrement is not transported in underground pipes and water certainly doesn't just pour out from a tap. Moreover, it is easy to forget the novelty of piped WS. These words of the Swedish author August Strindberg from a letter in 1883 could be a reminder of the fascination modern infrastructure aroused at the end of the nineteenth century:

> I met the most brilliant invention in Hamburg. There one crapped in something resembling a soup bowl, and when you looked around there was nothing to see, although you could swear that you had laid down a few meters, the dish was so clean after the service that you could eat genuine turtle soup out of it. (Jakobsson, 1999)

Almost 60 years later, in 1941, when infrastructural systems had become an integral part of city life, another author showed a similar admiration for infrastructure. It was Ludwig Nordström, journalist well known for a radio reportage during the 1930s, highlighting the poor hygiene standard in Sweden and coining the all but flattering term 'Dirt-Sweden' (Lortsverige). However, in this quote, he was more optimistic when it came to the benefits of infrasystems:

> Stockholm, like every modern city, is a wonderful creation of the soul, of imagination, calculation, inventiveness, dedication, sense of duty to help all people in the city to a reasonably human-worthy life. I see before me this wonderful creation of the human spirit in its entirety: first, the whole invisible city under the ground in the form of passages, drums, halls, machine halls, in which conduits of various kinds, silent and unknown to the general public, work and enable, for example, that it can quench its thirst with clean, bacteria-free water, can wash itself, shower, bathe, maintain a standard of cleanliness that makes it a small group in the great world of culture. I see the brilliant electricity plants, where the turbines whir as softly as cats purr, I see the coking out at Värtan, where the gas is produced, the telephone switchboard's serpentine tangle of cables, which allow all these Stockholmers to get in touch with each other in a second. (Blomkvist & Kaiser, 1998)

The research presented in this book is ultimately motivated by the many challenges facing Swedish and global infrastructure. Water and sewer installations, roads and streets are aging, and in many areas extensive maintenance is needed. Infrastructure in poor condition has significant adverse repercussions on the economy and environment alike. Furthermore, municipalities often lack support, resources, and capabilities to deal with the situation. This picture holds true also in a global perspective where urban systems for piped water, sanitation, roads, and streets have begun to see decline and disrepair even in high-income countries. These infrastructure sectors face large investment needs, especially in the face of global warming effects. The upcoming water crisis, partly because of a global over extraction and pollution, is described like this by Fishman (2011) blaming our reluctance to act on the 'brilliant invisibility' of water systems:

> Perhaps the most unsettling attitude we've begun to develop about water is a kind of disdain for the era we've just lived through. The very universal access to water that has been the core of our water philosophy for the last hundred years – the provision of clean, dependable tap water that created the golden age of water – that very principle has turned on its head. The *brilliant invisibility* of our water system has become its most significant vulnerability. That invisibility makes it difficult for people to understand the effort and money required to sustain a system that has been in place for decades but has in fact been quietly corroding from decades of neglect.

Furthermore, infrastructural systems often suffer from innovation deficit and are hampered by inconsistent and complex institutional frameworks.

In addition, large-scale infrastructures typically exhibit technical and institutional *inertia* and *path dependency* which means that decisions made during the design and construction phases affect the whole system for a very long time. The historical choices on technology and organization are difficult to affect after the systems have been built. It is also very expensive to change infrastructural systems due to the large investments already made (sunk costs). These factors make infrasystems conservative and the uptake of innovations slow. Earlier research (Blomkvist *et al.*, 2019) clearly shows that new innovations must be aligned with existing technical, organizational as well as socio-technical contexts to have an impact. This points to more fundamental, structural, and even historically grounded problems in infrastructure that need to be understood to fast-track a transformation toward sustainability. To put it bluntly, innovation-driven transformations of infrasystems are crucial to mitigate the upcoming climate crisis. As mentioned, I hope that this book will shed light on the historical legacy in infrastructure and reveal factors that need to be considered to reach an innovative and sustainable management of streets, water and sanitation.

1.2 THE ALLURING LEGACY OF ROME

When dealing with the history of municipal infrastructure, it is impossible to avoid ancient history and especially Roman achievements because Rome has a certain allure to historians of infrasystems. As indicated above, roads, water provision and sanitation, in one form or the other, have been around since the dawn of humankind. In historical accounts, researchers often start by a recapitulation of their ancient and Roman history, giving the impression, perhaps unintentionally, that we see an unbroken system evolution over several millennia, especially in water and sanitation history. In this book, I claim that ancient history does not matter very much, at least not in any concrete sense, in the development of Swedish or western municipal infrastructures. Even if the technology in these systems was well known and the envy of every municipal engineer, very little speaks for the existence of a technical trajectory stretching over thousands of years, at least in WS. Accordingly, the book discusses ancient and Roman history, not to paint a picture of direct heritage, but rather as a way of showing the *broken* trajectory after the fall of Rome until the reawakening of municipal infrastructure in Europe in the beginning of the 1800s. However, even if Roman technology and organization had little concrete impact, the reputation of Rome and to some extent Roman water law still affected the articulation of publicness in WS. It must also be noted that Roman and ancient water technologies did not disappear completely but 'hibernated' in some localities. In medieval times, monasteries and royal castles were quite often equipped with running water. One example is the castle in Turku (Åbo) in Finland, which was a part of Sweden at that time, built in the 1280s and equipped with pipes from a water well that is still functional today (Juuti *et al.*, 2009).

The second allure of Rome, apart from a belief in the existence of a technical trajectory, lies in the perceived connection between ancient infrastructure building, especially in WS, and *Bonum commune*, the common good. *Bonum*

Commune Communitatis or 'general welfare' refers to what benefits a society, as opposed to *Bonum Commune Hominis*, which refers to what is good for an individual. One example of this is Bjur (1988) who starts his, in many ways, excellent historical investigation of 'the art of water building in Gothenburg for 200 years' by directly connecting the efforts in inrastructure development to Roman perceptions of the communal good. In the first sentence of the book, Bjur asks the rhetorical question why the acronym *SPQR* is engraved on every manhole cover in the streets of Rome and on the front of official buildings. The acronym stands for *Senatus Popules Que Romanun*, meaning *The Roman Senate and People* or *The Senate and People of Rome*, and is an emblematic phrase referring to the government of the ancient Roman Republic. Bjur states that the engraving on the lids to the underground water and sanitation systems shows that they were built for the *Bonum Commune Communitatis*: 'The art of water building showed early on its quality of serving the common good.' Bjur is of course not wrong. The Roman art of water building was surely in many ways connected to *Bonum Comune Communitatis* and served the senate and the people of the city (*SPQR*).

However, the referral to the acronym *SPQR* and to an ideology of *Bonum Comune Communitatis* is still a bit misleading. First, the many references to these concepts originally appeared at a time in the Roman empire when the old republic had been replaced by autocratic rule under the emperor Augustus (27 BC to 14 AD). For example, coins with the inscription SPQR began to appear at the same time as Augustus attempted to legitimate his claims to have 'restored' the republic. Ancient historians claim that the usage of the acronym served to justify autocracy by connecting Augustus's rule to an earlier golden age to preserve the myth that the Republic still lived on. Since the time of Augustus, SPQR and this biased vision of Rome have repeatedly been used to connect various ideologies to the mythological power of the Roman republic. This rhetorical connection was, for example, apparent in the Italian Fascist movement under Mussolini. After the proclamation of the dictatorship in 1925, Mussolini's appropriation of Roman symbols was evident and functioned as a strategy to build legitimacy. Ancient Rome was portrayed as a model for political and military organization and as a symbol of Italian unity. Furthermore, to reconnect to WS, it was in fact Mussolini who popularized the use of the SPQR inscription on the manhole covers which has been interpreted as an ancient and true Roman symbol of *Bonum Comune Communitatis* (Benes, 2009; Hardwick, 2003).

To sum up, and as will be elaborated later, the many references to the Roman era and its alleged publicness in the history of Swedish and European pre-modern and modern WS were mostly rhetorical ornaments. By appropriating the grandeur of Roman technology and its reputation as a true communal good, advocates of public water and sanitation could motivate their proposals and tap into the glory of ancient Rome. Thus, Roman examples surely had an impact on modern system building although not in a concrete technical and organizational sense. Roman influence was more symbolic in the articulation of publicness in water and sanitation infrastructure.

1.3 PUBLIC AND PRIVATE GOODS

Having said the above on the rhetoric of a Roman legacy, it is still important to note that the history of public involvement in streets, water and sanitation is related to the debate on *Public Goods* among economists and political philosophers (also called *common* goods). The discussion on what constitutes a public or a private good has been ongoing since Aristotle and was, as has been indicated, an important part of Roman political life. The term *Res Publica* vs. *Res Privata* (public vs. private property/affair) is a living part of our Roman legacy.

A private good is something that a single individual can consume and by doing so prevents consumption of other individuals (*rivalrous* goods) or a good which is *excludable*, meaning that it is possible to prevent others to consume it, to 'draw a fence around it.' Food that we eat is considered a rivalrous good while listening to radio music normally is non-rivalrous. Landownership is considered excludable while streetlight and the air we breathe are non-excludable (Stanford Encyclopedia of Philosophy). Thus, public goods are, as opposed to private, what is shared by all, or many, members in a community and it is easy to see that an infrastructural system most often fits in this category. Furthermore, public goods, as infrastructure, are difficult for an individual or a small group to realize. The investment to build a road network or a system for WS is huge and there are big financial risks associated with getting return on the investment. The services delivered are not easy to price and the market is often uncertain before the infrasystems are fully operational.

These characteristics of public goods (and infrasystems) have led to a debate on *economies of scale* and whether they should be provided by the market or by the state and if they are so-called *natural monopolies* or not. This discussion leads too far. It is sufficient to ascertain that what Kaijser (1994) calls the 'Swedish model' for provision of infrastructural services most often included high state involvement by authorities or state-controlled monopolistic companies. As will be discussed below, state involvement was stronger in roads and streets than in WS. Nevertheless, the water and sanitation system surely has similar characteristics of economies of scale and natural monopolies as other infrasystems.

Discussions about public goods are also closely related to questions on whether they should be considered a *human right*. That is, if they are so important for societal well-being, they ought to be provided for free or for a minimum cost giving everyone the right to share the common resource. This issue is also related to deliberations on financing through individual tariffs or a general tax shared by all. Without going deeper into theories on public goods, it is sufficient to say that modern infrastructure for streets and roads, water and sanitation, bears a strong *resemblance* to public goods. But this has not always been the case for all three of them. I will return to the question on public and private goods and the issue of publicness in the end.

Furthermore, the discussions touched upon above by philosophers and economists are quite normative and include an ambition to define intrinsic characteristics of public and private goods and methods on how to manage them. I rather look at how these areas of essential human needs have been historically defined over time. I do not stipulate their nature, but instead investigate the *articulation* of publicness: how streets, water and sanitation have gradually evolved into our present day publicly managed infrastructural systems and thereby, as an end-result, turned into public goods-like resources. Another way of putting it is that instead of discussing whether these resources *really* are private or public goods, I ask the question if they have been articulated as a public or a private affair.

1.4 PURPOSE AND THEORETICAL INSPIRATION

There has always been some sort of service arrangement to provide for roads and streets, drinking water and sanitation but they have not always been in the form of infrasystems. The purpose of this book is to analyze how these arrangements have changed over time focusing on the tension between public and private responsibilities from the nineteenth century up until today. The general aim is to get a better understanding of how history affects present-day management of municipal infrastructure.

For a service arrangement to become a public responsibility, it needs to be *articulated* as such. As touched upon above, *articulating publicness* is a process whereby an area in society which earlier has been defined as being part of the private sphere, a task for the individual citizen or the household, is transformed into a task where public bodies such as the state or the municipality carry the responsibility. Furthermore, the level of publicness is not a given based on intrinsic properties in the resource or service in question, whether it is a public or private good in any definitive sense. Publicness is politically constructed and dependent on the actual historical context. The articulation of publicness means that areas considered as *Res Privata* are turned into *Res Publica*.

Following from this, I argue that the articulation of publicness includes two interconnected elements. First, a discussion on whether a certain area belongs to the private or the public domain at all. Second, after public responsibility has been ascertained, a discussion on what type of actor should be the performer of the tasks now declared public. However, the articulation process is not static, and it is not decided once and for all that an area belongs to the public sphere. I follow historians Hallenberg and Linnarsson (2016) in their analysis of a '... successive articulation of the public sphere ... (they investigate) ... the role of political discourse in articulating a stronger sense of publicness.'

Thus, I investigate the historical evolution of three municipal service arrangements in relation to *publicness*. I want to find out in what way these arrangements have been *articulated* as public or private concerns and obligations and how the perception of publicness has changed over time. I propose that current management in municipal streets, water, and sewage is strongly influenced by the historical development of whether these service arrangements should be defined as public or private.

Hallenberg and Linnarsson (2017) describe the concept of *publicness* as a multifaceted phenomenon: 'ideas about the public/common good.' The concept includes several different components:

- Who constitutes the community, who are included as 'citizens' and who are excluded?

- What resources, tasks, and activities are considered public, and must be carried out to ensure the continued existence of the entire community?
- Who shall carry out the above tasks, and who may act in the name of the public?
- What purposes, values, and ideological goals do the community identify with in connection to the public good in question (order, health, equal distribution, etc.)?

Hallenberg and Linnarsson have mainly focused on point no. 3 in the list above. They analyze a central component in the articulation of publicness, namely the ownership of different public arrangements: Who were the actors: private (commercial companies) or municipal bodies? In other words, they are mainly interested in the second element of the articulation process as I describe it above. In my analysis of publicness in streets, water and sanitation, the question of ownership is included, but it is not in the center of investigation. I also include the first element: the discussion on whether a certain area should be considered private or public at all.

To avoid misunderstandings it must be noted that I use the term *private* in a different way than Hallenberg and Linnarsson. They analyze the political debate on which actors were best suited to execute public tasks: public bodies such as municipalities or *private* actors such as commercial companies. This debate is very much alive today in Sweden and in Europe, where for example public domains such as health care and schools are run by commercial companies. In this book, *private* refers to private individuals and citizens (or groups) without commercial interests because in the history of streets and WS, the presence of commercial companies as owners has not been so prominent (but there are exceptions), and because I am interested in the process where these areas were articulated as public in the first place.

To analyze the articulation of publicness in streets and WS, I relate their transformation into infrastructural systems to several important contextual factors in Swedish history. But it must be noted that although I focus on Sweden, these contextual factors were in many ways also present in most European and North American municipalities at the time. The history of municipal infrastructure in Sweden is part of a general development in the whole Western world.

First, I discuss how three *general contextual* factors have affected the articulation of publicness in all the sectors. The first one relates to a long history in Sweden stretching back to at least Medieval times while the rest of the factors mainly relates to the nineteenth century. The general contextual factors are:

- The strong tradition of local/municipal self-governance.
- Urbanization, changing demography, and the industrialization process.
- The municipal reform of 1862.

I will also analyze the articulation of publicness in WS in relation to some *specific* contextual factors exclusively affecting the development in water and sanitation. These specific factors are:

- The divide between the private and the public.
- The social issue (concern for, and fear of, the working class and the poor).
- High mortality, cholera epidemics and new perceptions of health and sickness.
- The Sanitary movement and the Health Act of 1874 (which was part of the municipal reform).

In the chapter on water and sanitation systems in the twentieth century, two additional contextual factors are highlighted, which of course to some degree also influenced roads and streets:

- Environmental issues and water pollution.
- Sustainability, reuse, and the circular society.

Thus, publicness is articulated by various actors influenced by different time-dependent contextual factors. I will use these contextual factors to analyze how publicness has been articulated. At any given time, the community under investigation can be for example the village (with its village council), the town, the municipality, or the state. Furthermore, different resources, tasks, and activities are considered public in different historical settings and various actors are assigned the responsibility to carry out these public matters. The areas considered public are motivated and justified by reference to certain purposes, values, and ideological goals connected to the prevailing historical context. In the final part, when comparing publicness in these three sectors, I return to the issue of ownership in municipal infrastructure and how the building and operations have been financed.

As mentioned, the general aim of this investigation is to understand how history affects present-day management of municipal infrastructure. In other words, I am interested in historical *legacy*. In research on infrasystems, this legacy is often depicted using concepts from the field of large technical systems (LTS) (Hughes, 1987, 1988). In LTS, the already mentioned terms *inertia* and *path dependence* are used to describe the historical legacy which typically affects the system. Inertia, borrowed from classical physics and related to the mass of an infrasystem, is used as a metaphor to describe resistance to change (Nilsson, 2011). Inertia, in turn, creates path dependence, that is earlier choices in system design (technical and organizational) affect future development. In short, infrasystems have a large mass, and they are conservative and hard to change. In the following I will use these, and some other LTS inspired concepts when discussing certain *systemic characteristics* have influenced the shaping of municipal streets and WS. For example, I use the terms *system builder* and *systems culture* when discussing central actors in the development of municipal infrastructure (Blomkvist & Kaiser, 1998; Kaijser, 1994). These concepts and a few more will be explained further on. I argue that differing systemic characteristics can explain differences in the articulation of publicness in roads and streets and WS. However, it must be noted that I only cover some of the most important systemic characteristics. I have not ventured for a comprehensive system diagnosis.

1.5 METHODS USED AND LAYOUT OF THE BOOK

The investigation is based on a traditional combination of an extensive literature review of research from a wide range of various fields and a reading of historical primary sources. My findings and conclusions have continuously been validated in discussions with present day actors and historical experts (see above in acknowledgements).

The literature review, which is the biggest part, has resulted in a synthetization of previous research and my main contribution lies in the historical comparison of infrastructural systems that normally are dealt with separately. I have reinterpreted earlier results and brought together research areas that not so often communicate. The comparative and long-term perspective allows me to discover similarities and differences in the development of arrangements around streets and WS. Of these reasons, I argue that my literature review represents more than a regular state of the art in these fields.

The use of primary sources is centered on official documents, reports, and state lead investigations. I have not gone deep into political debates in, for example, various city councils or parliamentary disputes on infrastructural issues. This limitation in my research makes it difficult to present a fine-grained analysis of how publicness has been articulated through an investigation of the actors in various municipal authorities and organizations. I strongly advice for further research in this direction due to the strong local character of municipal infrastructure.

Parts of the book, concerning public and civic road keeping, are based on my earlier research and I occasionally use slightly revised versions of my own texts (Blomkvist, 2001, 2004). I have adapted them to fit in this project and translated some of them from Swedish. The same goes for quotes from primary sources and literature as well as titles in Swedish: all translations are made by me.

The method also includes what can be called an *historical contextualization* where I investigate the most important contextual factors mentioned above that influenced the development toward public infrastructure from the beginning of the nineteenth century. I have used research in medical, social, economic, technological, and political history highlighting the most important contextual factors in society at large that affected the *systematization* of streets and WS.

1.5.1 Chapter layout

After this introduction, Chapter 2 deals with three general contextual factors that affected the articulation of publicness in municipal infrastructure. Chapter 3 is about pre-modern and modern road and street history in relation to these contextual factors, and Chapter 4 includes a discussion of carriers of technology in roads and streets. Chapter 5 deals with pre-modern water and sanitation. Chapter 6 contains the specific contextual factors mainly influencing modern WS. In Chapter 7, the modern water and sanitation history is presented, followed by twentieth century water and sanitation in Chapter 8, including two additional contextual factors (environmental issues and sustainability). Chapter 9 is a discussion of carriers of technology in WS. Chapter 10 is a summative comparative analysis of streets and WS and a discussion on historical legacy.

doi: 10.2166/9781789063981_0013

Chapter 2

General contextual factors in the history of municipal infrastructure

To understand the articulation of publicness in municipal infrastructure of streets, water and sanitation, it is important to be aware of three important *general* contextual factors in Swedish society. They all affected the three service arrangements and infrasystems. Later, I will account for some *specific* contextual factors which mainly affected water and sanitation.

2.1 LOCAL AND MUNICIPAL SELF-GOVERNANCE

One fundamental factor in the history of these three modes of service arrangements and infrastructures has been *municipal independence*. Even if the term is a bit anachronistic before the municipal reform of 1862, local self-government was very strong and had been so for a long time, in both towns and in the countryside and a tradition in Sweden traceable to at least back to Christianization and the establishment of parishes in the Middle Ages. People living next to and attending the same church gradually developed local self-government to arrange their common affairs. Thus, local self-government was created from below and not through government order and the local community is older than the state (Nilsson & Forsell, 2013). Consequently, conditions in early modern Sweden were influenced by old structures of the local community with independent parish assemblies, where demands from the state were negotiated, creating a political culture based on peaceful solutions to conflicts which had the acceptance by the state. Swedish rule included a delegated *communalistic* principle which, in exchange for the support of the population, gave a certain degree of self-determination and forced the central power into responsiveness. This view is put forward in German context by historian Peter Blickle coining the concept of *communalism* in his research on political action by ordinary people in medieval and early modern Europe and within what sort of political framework they acted: 'Communalism is not an abstract term for just any form of commune, but rather for politically constituted communes equipped with such basics as legislative, jurisdictional, and penal authority. In this sense not only cities but also villages are an expression of the *societas civilis cum imperio* (civil society with government), to use the terminology of Old Europe.' I dare to say that the Swedish history of roads and streets and WS shows a high level of local independence, that is, communalism, in relation to the state and municipal authorities.

2.2 URBANIZATION, DEMOGRAPHY, AND INDUSTRIALIZATION

Urbanization was a relatively late phenomenon in Sweden. Around 1840, only about 10% of the population lived in towns and densely populated places. In the 1890s, 25% lived in towns and by the turn of the century this figure was 32%, but it was not until the 1940s that half of the population lived in (still small) cities. However, even though urbanization was quite slow compared to the European continent, Stockholm's population, for example, more than tripled, from 90 000 inhabitants to 300 000 between 1850 and 1900. In the first half of the nineteenth century, about three quarters of the population worked in agriculture. While the population increased from 2.3 million in 1800 to 3.5 million in 1850, the urban share was almost constant; it increased from 9.8% in 1800 to 10.1% in 1850. At that time, 70% of the towns had less than 3000 inhabitants and only 5% had more than 10,000. Stockholm was by far the largest with its 93,000 inhabitants; Gothenburg had 26,000; Norrkoping, 17,000; Karlskrona, 14,000 and Malmö had 13,000 inhabitants. Most towns had a distinct rural character well into the 1830s and the urban population produced around half the food they consumed (Blomkvist, 2023a). Nevertheless, the urbanization process affected water provision and sanitation profoundly. With higher population density, the problems with sewage and excrement management obviously worsened and the need for water, for street cleaning, fire protection and drinking, was accentuated.

Industrialization was a latecomer in Sweden compared to forerunners like England and Germany, but from the beginning of the nineteenth century, and especially from mid-century, the economic structure changed radically. Sweden was transformed from a poor country based on agriculture to an industrial nation. The first half of the century saw an early industrialization in agriculture, reforms in land ownership and new crops. Thus the industrial revolution was accompanied by an *agrarian revolution* which included new methods for more efficient farming such as the iron plow pulled by a horse instead of oxen, crop rotation, a focus on exports of oats and butter and better and larger livestock through conscious breeding and better fodder (Wiking-Faria, 2009).

By 1850 and onwards, Sweden went from being a raw material exporter to a substantial industrial producer of sawn timber and refined workshop products. Between 1890 and 1930, the modern industrial society took off. The gross national product (GDP) grew by 1.4% per year from 1800 to 1850; 2.4% between 1850 and 1890 and by 2.8% per year from 1890 to 1930. This translates into an increase of GDP growth per capita from 1.4% per year to 2.8%, even though the population increased substantially, which indicates a strong productivity development. From the very beginning, Sweden got most of its income from exports and the trade surplus has been around 4% per year (mean value) since the middle of the century up to the present. However, export-pull is not the only explanation. Sweden also had a substantial 'middle class' of self-supporting farmers forming a basis for a strong home market, a relatively even distribution of wealth and a population that could read and creative industrialists starting innovation-based industries such as Alfa Laval, ASEA, and Ericson in the 1880s (when Stockholm became the most telephone dense city in the world). Between 1840 and 1870, political reforms aiming at liberalization of the economy lifted trade tariffs promoting free trade, the establishment of a new law on joint-stock companies, the abolishment of guilds and the freedom for business (1846). The following period, 1890–1910, is often referred to as the era of 'Organized capitalism' when workers as well as industrialists started to organize themselves in trade unions and employer organizations and the Social democratic party was formed in 1889. This period was the origin of the famous 'Swedish model' for organizing economic life and the labor market (Blomkvist, 2023a).

There is one important area in this general industrialization process which I will come back to later: the development of technology and technical expertise. Technology advances were of course pivotal for the development of infrastructural systems. Watt's steam engine and later diesel engines and electricity were used in road construction, in digging for water and sewage pipes and for pumps transporting both drinking water and excrement in the pipes. Other examples are cast iron pipes with socket joints that were sealed with cast lead, in 1827, the use of iron for water mains was made compulsory in England, and new technology to filter water with slow sand filters, originating in England around 1830. Water quality improvement came slowly and in many cases water came from polluted rivers and lakes. Technological aspects in Sweden also included the training of civil engineers specializing in roads and WS and the establishment of private consultancy companies and engineering associations (and lobby groups) such as the Swedish Association of Municipal Engineers (1902), The Royal Automobile Club (1903) and the Swedish Road Federation (1914). They all became influential in propaganda and in setting standards and developing technology for roads and streets, water and sanitation.

Industrialization also meant that more people moved into towns and problems with bad housing increased with veritable slums in the bigger cities. The situation for the urban working class and for the poor worsened and the sanitary conditions, especially in towns, became truly appalling for many people. The so-called 'social question' was put high on the agenda which led to a strong public interest in WS systems (more on this later).

Roads and streets were of course also affected by the industrialization process, which will be discussed in more detail later. The effect can, with an expression taken from historian Jakobsson (1996), be characterized as the 'industrialization of roads.' Jakobsson studies how the natural water flows in Swedish rivers were transformed into a flow directed and controlled by an industrial rationality: the 'industrialization of rivers.' Roads and streets were not 'natural' in the same broad sense as the Swedish rivers. But after all, they had been there for a very long time and were an integrated part of the landscape and social structure. For modern engineers, the old road network was the same problematic entity as the rivers. It was all about taming, redirecting, controlling, and strengthening to make roads a part of the industrialized (and motorized) future.

2.3 THE MUNICIPAL REFORM OF 1862

The municipal reform of 1862 and its following statues must be seen in the light of the emergence of a more pro-active state in the second half of the nineteenth century. Kilander (1991) analyzes the new role played by the state from the end of the nineteenth century when state interventions became more frequent than before. Earlier liberals during the 1800s could combine a belief in the night watchman state and a minimum of state interventions, with an open mind toward state involvement in, for example, state-owned railway lines. The division in responsibilities was not between the *state* and the *private* but between the public and the private *interest*. The state had the right and duty to regulate what concerned the public interest but could not intervene in what was regarded as a private sphere. If the issue at hand only affected the lives of individuals or groups, nonintervention applied. But this perception changed and the state, through the Municipal Act of 1862, gave the municipalities the obligation and the tools to intervene in the private sphere. However, the municipal independence was still intact, and as will be discussed later, the state at the same time gave the municipalities a lot of freedom to decide on how much and in what areas to intervene.

Before the middle of the nineteenth century, the government in Swedish towns was upheld by the Burghers, and to some extent the property owners. The Burghers were citizens in towns that had the monopoly to carry on business, trade, and crafts (Burghership) and were organized in guilds (skrån). The Burghers had the right to participate in city decisions and they paid tax. During the Middle Ages and into the middle of the nineteenth century, the Burghers were part of the so-called Estate society where power was shared between the king and the four estates: Nobility, Priests, Burghers, and Farmers. Earlier the Burghers made up around 22% of the city population but this number fell to approximately 14% in the 1830s. The demise of the estate society and the more liberal trade legislations in the end of the eighteenth century and onwards meant that the original idea behind town privileges, the exclusive commerce rights (monopoly) given to the Burghers, gradually faded away. The estate society was formally abolished in 1866 (see below).

The most important town authority was the Magistrate which functioned as the city court and as an administrative body dealing with the day-to-day management of city affairs. The Magistrate was led by the mayor and several magistrates were (varying according to town size) appointed by the Burghers and the property owners. The other important authority in the towns was the Elders Council of the Burghers which represented the interests of tradesmen and craftsmen. A special body, which eventually became the most powerful authority, was the economic commission (Drästelkommission) dealing with the town budget, income, expenditures, and tax collection, a city department of finance. The Burghers most often controlled this organization by their majority of representatives. The third power center was the Church parishes (församlingar) which had their own parish meetings to decide on town affairs, taxes, and expenditures. In practice, these three had to agree on all important, and costly, matters which made the town hard to manage effectively.

Furthermore, the political leadership and the administration in Swedish towns were scattered before the municipal reform in 1862. Basically each town decided on their own, and it is hard to talk about a general governance model. However, the low level of unity was mainly due to a conflict between two logics. The first was the medieval praxis building on the estate society with the magistrates and the Burghers, in which the latter held most of the power. The other logic was, as mentioned, based on the ecclesiastical organization building on the parishes. In Stockholm, for example, with its eight parishes, all projects had to be approved by all. To deal with these problems, a special parish board was established to work out compromises between different interests. It seems clear that the hardships of getting unanimous decisions in, for example, infrastructural projects were a strong reason for the municipal reform. After 1862, the power was concentrated in one central body, the mandatory municipal council. It must be noted that in Sweden, as well as in other parts of Europe, the working classes did not have any representation in the municipal government based on the reform of 1862. For example, it was not until 1903 that the first representative of the working class became a member of the Stockholm city council (Juuti *et al.*, 2009).

The many stakeholders spurred administrative innovations long before the municipal reform. Already in 1811, the Burghers of Stockholm petitioned that the town finances should be transferred to a special commission jointly managed by the magistrates and the Burghers. The suggestion resulted in the establishment of the already mentioned *Drätselkommissionen*, a department of finance at town level, in which the Burghers secured a majority of seats. This move was of course aimed at breaking the frequent paralyzing power struggles between Burghers, magistrates, and the parishes that often made decisions on infrastructure investments and similar projects extremely difficult and unwieldy. The reason for these problems is quite easy to understand. When building an infrastructure in, for example, gas or water provision, it is impossible to connect all inhabitants in a town momentarily. Thus, people in the peripheral parts were forced to wait some time, often years, before they could be connected to the grid. Nevertheless, they were still obliged to pay for the project. This typical feature in infrastructure development caused envy and suspicion: why should we pay for services benefiting others? Furthermore, these problems often got worse because the people first connected to the grid were wealthy property owners in the city centers.

The number of Swedish towns and parishes, which in 1862 basically were transformed into municipalities, was almost 2500. The following municipal reforms of the late nineteenth century and in the twentieth century greatly reduced this number. As mentioned, Sweden now has 290 municipalities. In May 1862, Stockholm received its own municipal ordinance. It was common that capitals got a special position because the government (and the King) wanted to control the capital towns, where the national political power was concentrated. But already in 1817, the parishes had received the right of taxation in certain matters, while in 1862 taxation rights were extended to all the areas that did not belong to the state's obligations. Thanks to the municipal tax, the municipalities were able to take responsibility for the infrastructure that came with industrialization and increased population. From a managerial point of view, the municipality was regarded roughly as a joint-stock company where the shareholders' degree of decision-making was determined according to each one's share. For most of the municipalities, taxation became the most important source of income. During the 1870s, cities and towns together received on average just over a third of their total income via tax funds. That percentage rose to over 40% in the early 1910s. The second most important income soon became fees charged for gaslight, water, and electricity and the third income source was fees for selling and dispensing spirits and profits from municipal spirits companies. In the 1870s, an average of 15% of municipal income came from alcohol (Nilsson & Forsell, 2013).

Through the municipal reform of 1862, new principles for municipal administration had been created and municipal self-government had been strengthened. The Municipal Act gave each municipality the right to take care of its own affairs. In the preparatory work for the new municipal law, it was said that municipalities were not allowed to run for-profit companies. However, fees for the operation could be charged. The public undertakings that were run as companies were gasworks, electricity works, tramways, bathhouses, theaters, and in some cases hotels. Following Kaijser (1986), one can see a clear connection between public infrastructure building and the municipal reform. In the beginning of the 1860s, there was a culmination of municipal gas works and the beginning of municipal piped water systems. In 1870, 18 out of 20 of the largest towns in Sweden had built a gas plant (12 of them were owned by the municipalities). As mentioned, starting in Stockholm 1861, water works were built, followed by Karlskrona 1864, Jönköping 1865 and Malmö 1866.

It is evident that the arguments in favor for municipal self-governance in, for example, gas and water were supported by a report from 1859 which outlined the principles for the municipal reform. The municipal area of competence was defined as 'common order and housekeeping concerns' and the committee stated that municipal responsibilities should be rooted in the common municipal interests. Thus, the old principles of self-governance and independence were reinforced and also the obligation for every municipality to take care of public affairs. Every municipality had to deal with these issues, but they could do so in a way that they saw fit (Kaijser, 1986).

Again, the most important factor concerning infrastructure was the expanded possibility for the municipalities to levy tax from all citizens. Earlier it was the Burghers and the property owners, in towns and the farmers on the countryside, that paid for most of infrastructure expenditures. Furthermore, given the new income from taxation, the municipalities were able to put up a stronger security toward banks and other financial institutions. The cities borrowed money to a greater extent than the rural municipalities which can be seen in expenses for interest which increased from 9–10% to 14–15% from mid-1870s to 1920 and it was often the second largest item of expenditure after infrastructure, comparable to schools and more than poor services. Borrowing, in turn, was largely due to infrastructural investments. Between 1880 and 1910, the towns' borrowing increased sevenfold (Nilsson & Forsell, 2013).

The Municipal Act of 1862 was soon followed by four so-called city statutes detailing the obligations of towns. They were the Statute of Order (1868), the Fire statutes (1874), the Building statutes (1874) and the Healthcare statutes (1874). All the statutes applied compulsorily in cities. The City Planning Act of 1907 is also often included in this list. Concerning building and city planning, the cities had responsibility for their own urban planning, which included street management, since the seventeenth century but in 1874 consequently, these areas were regulated in national legislations. A new building law was issued in 1931 and came into force in 1932 at the same time as the new Town Planning Act.

Already at the beginning of the twentieth century, the need for cooperation between the municipalities increased, and in 1908 the Swedish City Association was formed and in 1920 the Swedish Association of Municipalities. They were merged in 1968 under the name the Municipalities Association (Kommunförbundet). The need to cooperate was a consequence of the strong expansion of the public sector, managed by the municipalities. From the 1970s, the municipalities came to have a growing influence over physical planning, and up to today environmental issues have also been included as a central concern. The post-war period was marked by two major municipal reforms. The municipality reform in 1952 was the most important as the number of municipalities was reduced from 2281 to 816, and the municipal block-reform of 1964–1974 further reduced the number.

Around the same time as the municipal reform of 1862, Sweden also decided to change its principles for parliamentary representation. This led to the abolishment of the former estate representation and the replacement of the 1810 parliamentary order, by the new parliamentary order of 1866. The new order meant the creation of two separate chambers of parliament (Riksdag): the first chamber, elected by the county councils, and the second chamber, elected in direct popular elections (but not yet universal and equal suffrage). This new parliamentary order of course had a propound effect on all aspects of political life in Sweden. I will briefly come back to these issues later. However, the complicated interplay between the new parliament, the government, and the municipalities is an area too large to cover. Nevertheless, Gullberg (1998) argues that municipal construction and management of various infrastructural systems helped in pushing for democracy in the towns and eventually gave rise to equal and universal suffrage on both the local and national levels. The large investments needed and the slow spread of the networks, where wealthy people in the city centers got their connection before more peripheral users, which still had to pay, fueled an intense dissatisfaction creating a strong force in political mobilization for democracy.

Turning back to the municipal reform of 1862, it is perfectly clear that it had an enormous impact on the articulating of publicness in WS as will be outlined later. Especially, the new Healthcare statute (1874) was important and of course the ability to levy taxes, and the extended possibility to take loans. Now began the era of public WS infrastructure. The impact on roads and streets was not that straightforward because in this area the state road authority already played the first fiddle. The municipalities and the reform of 1862 did not influence publicness in road and street keeping as much as in water and sanitation. The reason was also, which will be discussed later, the presence of inertia and path dependence. The traditional structure of the road sector was hard to change.

doi: 10.2166/9781789063981_0021

Chapter 3

Pre-modern and modern roads and streets

The road traffic system in Sweden has three administrative levels and it is not possible to write about municipal *streets* without mentioning the other two categories public (state) *roads* and *civic* (local) *roads*. Their history is intertwined and stretches back into medieval times. As already indicated, in the following, I use the terms *road* when talking about the local and state levels and *street* when talking about the municipal level. Some parts of the pre-modern and modern Swedish road history have been published earlier by me (Blomkvist, 2001, 2004, 2010). I have adapted the texts for the purpose of this book and translated them from Swedish and some references have been kept.

3.1 PUBLIC ROADS

All roads lead to Rome, as the saying goes. It illustrates the importance of roads for the emperor, or the sovereign's ability to control the territory. Julius Caesar held the prestigious position of 'curator of roads' before he became the Roman emperor. Since the days of Rome, the road's legal status has been governed by the principle of the public's right to free passage (right of way). But this right of way did not exclude road tolls or other fees. In Great Britain, road peace and free use of the public roads were legislated early in the Middle Ages but as shown in the legend of Robin Hood, the peace of the road was not always upheld. Nevertheless, 'The King's Highway' was supposed to be a place where travelers had the right to stay during movement protected by the Crown which replaced the landowner's rights of the road space. Maintenance 'in kind' by the farmers was the usual form of road tax throughout the world, although varying degrees of coercion occurred through day labor and work done by convicts (especially in the USA, so-called 'chain gangs'). Road building has often been closely linked to military motives. The Roman roads are perhaps the most famous example from antiquity. More recent examples are Napoleon's upgrading of the French road network, by the establishment of a national network of main roads ('les routes nationales') and legislation of the state's right of expropriation for road construction. Also, the Italian 'Autostrada' and Nazi Germany's 'Autobahn' had military strategic motives as advance corridors for the motorized infantry (Lay, 1999).

In Sweden, and since far back in history, we see connections between road management and the emergence of some form of centralized power apparatus. Publicness in road building was articulated as the interest of the chief, the king, or the state. From the Viking Age, we have runestones that testify to a rudimentary form of organized road building. Seen from legal historical perspective, road legislation carries a significant legacy that is perhaps unique compared to other areas of law. Rules for road management, already codified in Medieval laws, survived for a very long time, and still influence road building and maintenance today (Blomkvist, 2001, 2010).

In towns as well as in the countryside, public roads were managed as a public undertaking by the property owners as a tax payment in kind. This praxis was established during the Middle Ages and each farmer was given responsibility based on the agricultural land and share in the village. Rules for public road management were formally established in Magnus Eriksson's national law from the 1350s. Road maintenance was based on value and carrying capacity of the plot of land in the countryside, and for the city it was calculated in the same way, that is to say (roughly) what we today call a property tax. This way of managing public roads as a 'communion imposition' or 'inconvenience' (menighetsbesvär) was based on the principle of 'utility and interest', meaning that landowners living nearby were to manage the road because they were seen as the beneficiaries of the service delivered. Thus, public road regulations in Magnus Eriksson's national law were also applied in the cities and later transferred to the City Act, which also bore his name. Both laws formed the basis for Swedish legal administration in the road area, until they were replaced by the 1734 law. It is always tricky to write history according to laws and regulations and when it comes to road history of older times, we know very little about the concrete process of road building and maintenance. However, it seems clear that the state tried to manage the road sector with a gentle hand and that the goal was to reach amicable agreements. Without some *communalism* and the goodwill of the farmers and property owners, no roads were built or maintained.

The state's interest in public roads during the Middle Ages and up to the eighteenth century was essentially articulated as the central power's need to control the territory which was gradually complemented with a desire to promote trade and manufacture, both in the nation and in towns. King Gustav Vasa, in the sixteenth century, wanted to create a coherent network of roads adapted for wagons. The public roads which were called country roads, main roads, or royal roads certainly had a status as nationally important but were up until the middle of the seventeenth century not suitable for anything other than horse riding. During the seventeenth century, however, national roads became increasingly important for the crown and the state. Under Gustav II Adolf, the road network was significantly upgraded, mainly for military reasons. In 1636, four years after his death at the battlefield of Lützen, the guardian government of his daughter Kristina formed the Swedish Postal services. Queen Kristina followed up her father's road initiatives and in 1649 introduced rules to maintain passable roads between the inns and set up a national shuttle service. In the same year, she established a position as the kingdom's road master. Road construction and road maintenance were carried out by the king's officials (Landshövding) in the counties. Still, this practice was based on amicable agreements between the state and the land-owning congregation. But even though the road system was expanded and even though the roads became increasingly important, not much happened around basic legislation. The medieval organization was stable throughout the period and the principle of utility and interest prevailed. In the famous 1734 law, which was upheld, albeit with many additions, until the end of the nineteenth century, the state's interpretation of the interest and utility principle were kept in the legal text. But the 1734 law quickly became outdated. It came under severe pressure from the political reforms such as the Municipal Law in 1862 and the Constitutional reform of 1866 and also the gradual industrialization process.

The state's interest in the roads became even greater during the nineteenth century and more and more roads were defined as public roads. An example was a royal decree from 19 February 1824 regarding additions to the Building Code (BG 25:1) where several new public roads are enumerated to the so-called staple cities. The staple cities relied on trade and the increasing importance of industry. Furthermore, the articulation of publicness was also visible in the issue of expropriation of land for road purposes. Several regulations on expropriation were drawn up from the middle of the nineteenth century, which led to the Expropriation Ordinance in 1866 and the Expropriation Act from 1917. In 1930, the concept of 'right of way' was introduced, which was also applied in the cities. The right of way meant that the state or city was allowed to build a road over private land if required. During the nineteenth century, the road system was increasingly centralized and professionalized. In 1813, the Royal Committee for Road Construction in the Northern Provinces was established and in 1841 the Road and Water Works Board (VoV) was formed, and the country was divided into five cross-county road and water construction districts (i.e. water power for electricity generation). Through VoV, state subsidies began to be paid out for building new roads as well as strengthening and straightening of old roads. With this development, the articulation of publicness in roads was strengthened and public roads became a public responsibility to an even greater extent. In 1851, the Road and Water Works Board appointed road engineers in every district in the country. To get government funding, VoV's participation was required, and the technical experts therefore gained an ever-increasing influence over the road system: 'Thus began the professionalization of road construction and maintenance that would not only raise the quality of the roads, but also fully transfer them into the hands of (public) specialists a bit into the next century' (Pettersson, 1988).

In 1891, a new law on public road keeping was introduced. The organization of the public road system had been shown, medieval origins. It was based on an agrarian logic where small self-sufficient units were the starting point. As industrialization gained momentum at the end of the nineteenth century, this logic was perceived as overplayed and out of date. First, agriculture became more market-oriented and the need for transport outside the immediate area increased. Second, industries grew and required roads for its raw materials and products. The farmers who were obliged to maintain the roads came to see the road burden as unfair because the new industries did not have to pay for the roads. Third, the building of the railway network meant that the need for roads increased as goods and passengers had to travel to and from the stations. Fourth, the route of the railway meant that population and business were concentrated in connection with the nodes of the railway network. The Road Act of 1891 is usually said to have transferred road maintenance to the public domain, and what was new in this Sweden's first proper road law was that the responsibility for road maintenance passed from the agricultural property to a new legal entity, the so-called road district. The road districts, or the road municipalities as they were called, were based on the county or part of the county and the number was 379. A road board was appointed to lead each road district. Furthermore, a road tax was introduced to be paid in cash for groups that were previously excluded from road maintenance. Now industry could be made to pay for the roads. But even though the road district was now in charge of road maintenance, in-kind maintenance was not abolished. The farmers were not given the opportunity to pay the road tax in cash but would continue to maintain the roads according to the old way. It is striking how the law tried to balance two different logics: the agrarian, local with maintenance according to the utility and interest principle, interpreted as agriculture still having the greatest benefit from the roads, respectively, the industrial and national logic, where the benefit of roads for industry was recognized. Many of the old problems in the road system were brought into the twentieth century. In retrospect, it is easy to agree with historian Pettersson (1988) when he writes:

> 'In principle, the road fund could have taken over at once … Admittedly, road maintenance had probably initially become somewhat more expensive and the tasks of the road boards more numerous, but the roads had been better maintained and that had undoubtedly simplified things considerably. As early as 1895, there was the apparatus required to take over all maintenance of the public roads through the road fund and road board, the problem was that they were also forced to keep the old methods.'

Criticism of the Road Act was strong. When the legislature in 1921 opened the possibility for the road board to take over the entire operation under its own auspices, there was room for a thorough structural change without the law needing to be rewritten. By the mid-1920s, more than half of all road districts had taken over, in 1928, 72 remained and in 1930 only 12 districts still followed the old in-kind principle (more on this later). In 1934, yet a new Road Act was introduced, the most important change was the formal abolition of road keeping in-kind. Something that *de facto* had already happened. The road districts under the leadership of the Swedish Road and Water Administration became responsible for public roads in the countryside, and the towns were to cater for public roads within its borders. Streets were still their responsibility. In 1944, the public roads in Sweden were nationalized and the utility and interest principle moved up to a societal level. Now, it was the entire nation's joint obligation to keep roads and publicness in roads was articulated as a true national interest. The Road and Water Works Board (named the National Road Administration from 1967) were put in charge for the whole road (and motor traffic) system. From 1944, each county had its own road administration, but from 1992 several counties were merged into regional districts. In 1959, the *Road Plan for Sweden* was presented and adopted by a unanimous Parliament and the ground was paved for the entry of mass motorization. With the so-called *Road Plan 70*, the cities were also drawn into road planning adapted to the car (Blomkvist, 2001).

3.2 CIVIC ROADS

Internationally, civic roads are a unique Swedish (and Finish) road category which is managed directly by the actual road users living nearby. The laws regulating civic roads have a direct and unbroken heritage from the way they were managed in medieval villages by the land-owning farmers. During a period of 500 years, civic roads were the responsibility of the village council and the landowners using the road (Blomkvist, 2010).

The earlier national road regulations had, as noted, their center of gravity in public roads. The legislature wanted to regulate the construction and maintenance of the roads that were considered important to the state. But the laws also contained some statutes on smaller roads in the villages. From the thirteenth century, the law stated (Upplandslagen and Västgötalagen): 'If village men want to build roads, other than those that belong to the state, then one wants to build and the other doesn't, then the one who wants to build is given the right to do so and is backed up by the King's military and pledge' ('våld och vitsord'). Magnus Eriksson's national law, from around 1350 (and Kristoffer's national law in the 1450s), contained similar provisions: 'Now if farmers want to lay a road through the village, they may do so, if they among themselves agree.'

As mentioned, regulations of public road keeping were also applied to the farmers in the villages and they were required to build and maintain public roads and to provide ferries and bridges. For civic roads, the basic rule was that roadbuilding should be on uncultivated and common land, and those who had utility and interest in the road participated in its maintenance. The laws also established the right of a farmer to use someone else's land for an exit route if needed and if compensation in land was given. Thus, since a very long time, farmers have had the right to build a road if they could demonstrate the benefit. No property would risk ending up without the possibility of connection to rest of the road network. In these early laws, there was nothing stipulated about road width or technical quality. It was left to those involved to decide.

Already in the medieval laws, a very important difference between civic and public roads can be seen in terms of the relationship between the property's assessed value and the road's utility. For the public roads, the individual landowner's obligations in road keeping were related to the agricultural capacity in a direct way and the assessed value (like in today's property tax) was used as a proxy in the calculation of the road burden. It did not matter if the property had any real benefit from the road. If a road was designated as public and classified as a parish, district or county road, each landowner was automatically assigned a road lot. This was not the case for civic roads. Here, a more direct and user-oriented utility concept applied, which did not have as strong a connection to property value. Whoever could demonstrate the benefit of the road was allowed to build it and those who benefited from the road should be involved in the project, each according to the actual benefit the road gave to each. However, there is no information on exactly how these road lots were determined in the medieval sources, although it is likely that the agricultural capacity and share in the village were important here as well. This medieval variant of the utility and interest principle, exclusive to the civic roads, has survived in today's legislation. As will be discussed later, in the section dealing with pre-modern drinking water, a clear inspiration for the legislations on civic roads came from early water legislation. But these water laws did not target drinking water provision at all. The water laws were all about dikes, that is drainage of farming land, and later focusing on water issues such as hydropower (Blomkvist, 2010).

In the law of 1734, the division between civic and public roads became explicit for the first time in legal history. It regulated how civic and public road keeping were to be carried out through 'general impositions' (allmänna besvär). Furthermore, in the 1734 law, civic roads were not only linked only to the ownership of agricultural land. Even local mills and other common facilities such as summer farms (fäbodar) could be responsible for civic road keeping (Blomkvist & Larsson, 2013). But even though the state wanted civic roads to be managed only by those with the most clearly expressed interest, it seems to have been difficult to draw boundaries. As late as 1828, the state had to clarify in a royal letter that disputes about civic roads would be settled in courts as disputes between private individuals, not as administrative cases in the county court, according to the rules for public roads. A conclusion regarding the distribution of road responsibility between the civic and public roads in the 1734 law is that the legislator wanted to keep civic roads within the sphere of private law as far as possible. The state did indeed issue some regulations on road maintenance, but it was up to the local road managers to regulate the finer details in some form of joint agreement. In principle, not much changed regarding legislation for civic roads. The law of 1734 was complemented in new statues 1907, 1926, 1939 and finally in 1974. As mentioned, the utility and interest principle has survived and property owners living in proximity to the road have basically the same rights and obligations today.

However, after WW2 and the beginning of mass motorization (Blomkvist, 2010), civic roads were more tightly aligned with public road keeping. The state road administration appointed special road engineers to manage the interface between public and civic roads and took the initiative to form a special organization for civic road keepers, the Civic Roads Federation (Riksförbundet Enskilda Vägar, REV) founded in 1949 to coordinate the various local road associations. Today, REV organize around 13,000 of the 34,000 civic road organizations. Another interesting interface can be found between municipal street keeping and civic roads. Municipalities often own land and properties, and therefore become members of civic road associations. The municipality is obliged to participate and contribute to road maintenance of these civic roads to the same extent as other property owners. It is also common for municipalities to contribute to civic road keeping by financial contributions or by taking over the road management. A reason why the municipalities step in is an ambition to improve the standard of a certain road or to even out injustices when civic road keepers are double-taxed, as they already pay for municipal streets and public roads (Blomkvist, 2010; Blomkvist & Emanuel, 2020).

3.3 MUNICIPAL STREETS

Municipal street management has evolved in close relation to public and civic road keeping. As mentioned for a long time, street keeping was based on the same principles guiding public roads on the countryside and civic roads within the villages. The utility and interest principle meant that property owners living close to the street were responsible for both building and maintenance (and cleaning) for the street outside the property border stretching to half the width of the street. These rules were laid down in the 1350s, Magnus Eriksson's national and city law and repeated in the law of 1734. This meant that since the Middle Ages, property owners, both Burghers and others, were obliged to cater for street building and maintenance. Public *roads* passing through the town as well as parks and town squares were the responsibility of the city authorities and the Burghers. Outside the town border, the county and its property owners (the farmers) managed public roads.

With the increasing public road expansion during the nineteenth century and when towns grew bigger and traffic, internal and passing through, increased, public roads within the city limits were often included in municipal duties. However, according to Schalling (1932) before the municipal building statute in 1864, special legislation for towns were largely missing in national laws and the legislation on municipal streets were basically unchanged and based on older customs and can been found in rules for building and property development such as the various building and planning acts that were issued. Towns basically had the freedom to manage streets as they saw fit, and between different cities the distribution of street responsibilities came to differ where some cities had started to take over the road and street maintenance from the middle of the nineteenth century and financed this with general taxation. In some towns, the magistrates or the Burghers decided to manage both roads and streets as a public undertaking even earlier. In Gothenburg, the Burghers in 1811 appointed a Shuttle and Haulage board, which would construct and maintain roads. And in 1845, a special committee began to distribute the costs of street maintenance between house owners and the city's other residents. In Stockholm, a special administration was established for streets in 1845 and for gas lighting in 1850. In 1861, they were merged into the streets and lighting management board. It must be noted that these arrangements were introduced before the Municipal Act of 1862.

However, the Municipal Act, with its stronger power to levy taxes, gradually changed both road and street keeping in the cities. It seems probable that most towns had taken over both street and road management as a public undertaking in the beginning of the 1920s. Gradually, after the municipal reform of 1862, the principle of utility and interest was loosened and with the Town Planning Act in 1907, and stronger in the Town Planning Acts of 1931 and 1947, cities and city-like communities were obliged to provide for streets in areas with an established city plan. The legislation for urban planning legislation further stated that public roads included as a street in the city plan should belong to the city and thus towns took over the road maintenance for part of the public road network, an obligation which was kept in the 1934 General Road Act. Thus, the road legislation of 1934 codified a reality already existing when it stated that the new road district, and each town was its own road district, were responsible to, apart from managing city streets, maintain and in some instances build, public roads that passed through the town.

When the public roads were nationalized with the 1944 Road Act, public roads in the countryside belonged to the state while the public roads in the cities belonged to the cities. State road keeping was managed by the Swedish Roads and Waterworks Board and smaller communities were included in these rural areas. Also in 1944, a state subsidy was introduced to even out municipal costs due to increasing traffic demands in more densely populated areas. These rules were somewhat revised in the new Road Act of 1971 but the division of responsibility between the state and municipalities did not change much compared to the situation before the nationalization of public roads. In connection with municipal reforms of the 1960s and 1970s and because of changed settlement structures and travel patterns, the issue of increasing municipal responsibility for public roads was brought up. The idea was to harmonize legislation around public roads and municipal streets and that more cities should be involved in public road management which led to several investigations into the municipalities' obligations. For example, a state investigation in 1977 suggested that a new road category should be established: 'municipal public roads.' However, special national regulations on municipal public roads and streets never became a reality. Instead, street regulations were incorporated into the Planning and Construction Act (PBL) of 1987. In short, municipal self-determination and responsibility won over state control. Around 1980, 110 municipalities were road managers for public roads and since 1987, the number of municipal road managers for public roads has gradually increased to include 206 in 2018. In 1992, the state took over the responsibility for the most important thoroughfares in cities and larger urban areas (Tällberg, 2018).

3.4 PATH DEPENDENCE IN ROADS AND STREETS

Road maintenance carried out by the farmers remained for a long time, despite attempts to articulate road maintenance as a public task performed by the authorities. As mentioned, this arrangement for public road keeping was since the Middle Ages until the 1920s, a tax payment in kind. Historian Lindberg (2022) shows that this extremely decentralized way of managing roads was unconventional in an international perspective. Only Sweden had an arrangement where both the financing and the execution of road management laid in the hands of the same local authority and thus outside the regular parish or municipal administration. This way of handling road issues gave the local road managers a very strong position. Furthermore, he successfully argues that despite state initiatives to build new roads during the nineteenth century, the Achilles heel of public road keeping was road maintenance in kind according to the utility and interest principle. Lindberg convincingly shows that traditional road keeping was tightly inwrought with other institutional arrangements in Swedish society and therefore extremely hard to change.

The Road Act of 1891 strengthened, despite contrary intentions, local self-government in the road sector. The newly formed road districts became an independent form of municipal administration, where voting rights remained graded according to income and wealth, and where companies also had voting rights. The problem was that within a road district, actors with many votes could control and block decisions in the board. Lindberg claims that municipal self-government in the road sector prevented fiscal justice between the municipalities through municipal tax equalization. During the latter part of the nineteenth century, the municipal tax pressure rose significantly and the question of an equalization of the tax burden between municipalities became a burning domestic political issue, since a large part of the municipalities' cost increases was due to government decisions that applied to all the country's municipalities, for example, infrastructure, hospitals, schools, and so on. Between the years 1900 and 1925, the municipalities' expenses rose seven times and the debts increased fivefold. And these increases were far from evenly distributed among municipalities. It was thus the reluctance to contribute to municipal tax equalization that caused a reform of the road sector to be delayed, although most people realized that in-kind maintenance had played out its part. It was difficult to reform road maintenance due to far-reaching individual and local self-determination, where many changes could be blocked by individuals or by local communities. The strong position of the municipal self-government basically prevented all attempts to improve road quality. The nationalization of the road network in 1944 was thus a drastic way to completely disconnect road maintenance from local interests to achieve tax equalization.

Following Lindberg, I would argue that it was *institutional* inertia and path dependence that slowed down the modernization of the Swedish public road sector. Although I agree with Lindberg, I believe that institutional inertia and path dependence were not as strong in municipal street management, especially in the larger towns. As I have shown, both city streets and public roads passing through the towns were in many cases incorporated in the obligations of the municipality since the first half of the nineteenth century. This expansion of municipal engagement in streets and roads was realized before the municipal reform of 1862 and the new road legislation of 1891. It is quite clear that towns had a high level of freedom in dealing with street and road issues and some cities had started to take over road and street maintenance from the middle of the nineteenth century, financed by taxation. Furthermore, as mentioned, in the larger towns such as Gothenburg and Stockholm, the magistrates or the

Burghers decided to manage both roads and streets as a public undertaking even earlier. When the Municipal Act was introduced, with its stronger power to levy taxes, it gradually moved both road and street keeping in the cities toward even stronger public engagement. Furthermore, this increased articulation of publicness was most certainly strengthened by the adoption of the new Building statutes in 1874 and it seems probable that most towns had taken over both street and road management as public undertakings in the beginning of the 1920s.

These last remarks are based on a synthesis of secondary sources and therefore preliminary. More research on primary sources is needed. Nevertheless, I claim that my results on road and street keeping in the Swedish towns complement Lindberg's results. It seems to me that the conservatism he finds in public roads in smaller municipalities on the countryside is not found in the larger cities. The old habits of street and roads management faded away faster in the towns, and they were able to modernize the sector earlier. In the towns, the strong municipal independence did not cripple road and street keeping as it did in the countryside. Instead, the city authorities used their strong self-governance to adapt the sector to modern demands.

3.5 THE ROAD AND STREET SYSTEM TODAY

As been discussed, the present-day Swedish road system is divided into three parts. State (public) *roads*, including so-called national roads, are managed by the state. Municipal streets are managed by the 290 municipalities (including towns). However, the municipalities' street administrations are obliged to follow standards and specifications given by the traffic authority which are published and updated regularly. The third level consists of *civic roads* which are managed directly by the actual road users living near the road, organized in around 34,000 local road associations. This part of the road system is run by individual property owners, but the state road administration oversees and controls technical specifications, and distributes state subsidies to these often small, but important *capillaries* of the grid. The present road network in Sweden includes approximately 100,000 km state roads, 40,000 km municipal streets, and 150,000 km civic roads (Blomkvist, 2010).

Public roads are primarily regulated by the Roads Act (Svensk Författningsamling (SFS) 1971:948) and the implementing regulations in the Roads Ordinance (SFS 2012:707). The current Road Act has been reworked in parts but has the same foundation as the Road Act from 1944, when public roads were nationalized. Since 1971, the law has been adapted to the Environmental Code and the new Planning and Building Act and to regulations for the national plan for transport infrastructure (SFS 2009:236) and county plans for regional transport infrastructure (SFS 1997:263). Technical requirements for roads and streets are issued by the Swedish Transport Administration in cooperation with the Swedish association for municipalities and counties (SKL). These requirements and advice for the design of roads and streets are binding for the state, while they are advisory for the municipalities.

Municipal streets are regulated in the Planning and Building Act (PBL) (SFS 2010:900) which broadly regulates the planning and construction of property through so-called 'detailed plans' which is a planning instrument with legal effect and the basis for a municipalities' planning monopoly. The PBL replaced previous city and building plans (Tällberg, 2018).

Rules for civic roads are regulated since 1997 in the Facility Act (Anläggningslagen; SFS 1973:1149) when the law on civic roads from 1939 was changed. The purpose of the change was to achieve a more modern and uniform legislation with simpler rules and added regulations on *community facilities* (Gemensamhetsanläggningar) with special rules for civic roads.

As related above, the three categories, public roads, streets, and civic roads, are based on different types of legislation. The Road Act with its rules for the right to build roads on private land (right of way) is the strongest law of them all. But despite this, the road network is a relatively well aligned and cohesive infrastructural system with its three integrated levels. In other words, and using a term from earlier research, the road and street sector exhibits a strong *vertical integration* with a distinct system builder controlling each level (Blomkvist & Larsson, 2013; Blomkvist & Nilsson, 2017). As will be discussed later, this strong vertical integration is different from water and sanitation where we, at the present, can see a development toward *horizontal integration* (Alm & Paulsson, 2023; Alm *et al.*, 2021).

doi: 10.2166/9781789063981_0033

Chapter 4

Carriers of technology and publicness in roads and streets

When roads and streets gradually became an integrated infrastructural system in the first decades of the twentieth century, modern technology was introduced by a new set of actors taking on the role of system builders or system promoters alongside the state road administration. These new actors changed the articulation of publicness away from traditional state interests such a territory and trade toward the building of a Swedish industrial welfare state and a car society. From now on, publicness was associated with future visions of modern technology, a motorized industry and with the dream of a private automobile for everyone. This new articulation of publicness started quite slowly and accelerated during the 1930s followed by a full-scale implementation after WWII. Borrowing a term coined by Edquist and Edquist (1979) and used by Jakobsson (1996) and Olsson (2000), I call these actors *carriers of technology and publicness*. I want to stress that these new actors were not alone in dreaming of a future car society. The dream was shared by the state road administration, by industry and political parties and not the least, by a vast majority of the Swedish people. As will be discussed later, the same type of actors was also present in WS.

The chapter starts with a discussion on technical development and the new actors promoting technology and industrial management models in the sector, followed by a discussion of the development of a certain systems culture in roads and street management.

4.1 SYSTEM BUILDERS AND TECHNICAL DEVELOPMENT

As has been discussed several times, the state had a clearly pronounced ambition to articulate *publicness* in the road sector for hundreds of years. Through the state road administration, gradually from the beginning of the nineteenth century, public roads, streets, and civic roads were transformed into a well aligned infrastructure, a modern infrasystem and the state had become a system builder.

For a very long time, road building and maintenance had been performed by landowning farmers using ordinary tools and equipment such as shovels, crowbars, pickaxes, and iron skewers. Material for the roadbed, gravel, stones, and sand was fetched in proximity to the road and transported with the farmers own horses and wagons. Road keeping was not seen as a prioritized task compared to actual farming. It was often executed by farmworkers when there was room in the schedule which resulted in irregular and uncoordinated efforts. There are many accounts from road travelers describing that one part of a road could be well kept while the next stretch was in a terrible state and not yet maintained by the farmer assigned for that part of the road. Road networks and similar gridded infrastructural systems are a concrete illustration of the old proverb 'a chain is only as strong as its weakest link' (Blomkvist, 2001). Road keeping as a tax payment in kind was perhaps sufficient until the industrial revolution picked up speed in the end of the nineteenth century and road transport became increasingly important. The new demands put on the road network changed the perception of the old methods and technologies. Old style road work was increasingly seen as an obstacle, or in the words of historian Hughes (1987), as a 'reverse salient' for systems development and expansion. When the private automobile appeared at the turn of the century, the critique was multiplied. The advent of the car also meant that new organizations stepped in articulating publicness in roads and streets with a clear connection to automobility. As will be discussed, these organizations used arguments based on industrial technology and management models to promote their case.

Swedish motorists were early to realize the importance of organization. The Royal Automobile Club (RAC) was founded in 1903, when only a handful of cars existed in Sweden. Car ownership was, and still was for many years to come, a privilege for the very rich, and the automobile was mainly a sports and recreation vehicle. The most important outward objective of the RAC was to change the deeply rooted view in the public mind of the car as a toy for rich boys and a menace to society, a definition of the car that was quite accurate, at least up until the mid-1920s. Directly inspired by the so-called Good Roads Movement in the USA, the Automobile Club took the initiative to form the Swedish Road Federation (SRF) in 1914. Joining the motorists were commercial interests in road building, local and regional road keepers and, most important, staff from the state road administration, county governors and members of the Royal Corps of civil engineers, all educated in road and water engineering and members of the Swedish Technology Association. The road federation was a truly *corporativist* organization where State and county officials joined forces with commercial interests and technical experts. This mix-up of actors from different areas in society was not uncommon in Sweden at the time. Currently however, a stronger emphasis is put on separating public and commercial interests, and state officials can no longer be members of commercial lobby groups (Blomkvist, 2001, 2004).

Already in its first year, in 1914, the SRF launched a course in 'rational road keeping' in Enebyberg, north of Stockholm. The audience were traditional road keepers, farmers, and property owners, acting under the prevailing regime of road management in kind, following the utility and interest principle. These courses were seen as the best means of agitation for the road federation and over the years many were held all over Sweden. Expectations were high and an ambitious lecture program was planned: 'The need for and advantage of good roads,' 'Rational Road construction,' 'Rational Road maintenance,' 'Road construction and road maintenance abroad,' 'The importance of wider wheel rims,' and 'Reforms in road legislation' were titles agreed upon. The main trust in the argument during these courses was against the reluctance from the farmers to accept rational and industrial road keeping using modern machines. It was stated that most farmers had the wrong valuation of time itself. They thoughtlessly made several trips, with little cargo in their carts. The road lobbyists hoped that the value of time in the industrialized society would pave the way for the new and modern man who realized the importance of roads. The automobile that was 'first acquired for purely practical reasons' would later help shaping this new attitude toward time and ultimately force the improvement of the roads. The fear of labor-saving road equipment was seen as a sign of backward-looking hostility to technology: 'The Swedish Road federation is on the right track, as it does not waste its energy on kneading the old, outdated sourdough, which is called in-kind road management … (with its) absurdly driven parceling of road maintenance' (Blomkvist, 2001, 2004).

To build rail and waterways was considered an artwork of scientific status by the civil engineering community of the day, while road building was not. Road keeping was done by peasants and not a job for the technical expert. This was something that the road engineers wanted to change. They had a strong incentive to expand their professional field of expertise and to force the traditional road keepers out of the market. The SRF became the key professional organization for the road engineers in their struggle to gain society's acceptance of their claim on expertise and monopoly in the road sector. To join up with the engineering community was a perfect strategy, from the motorists and the commercial interest's point of view. Through the SRF, they could argue for better roads adjusted to the automobile using the allegedly neutral and scientifically objective arguments of the civil engineers.

To make a long story short, the SRF managed to, metaphorically speaking, *industrialize* road keeping. The old tradition of work in kind was abandoned in the end of the 1920s, the engineers became uncontested experts, technical rationality was applied, industrial methods and machinery was adopted and the revenues from the introduced car and fuel tax (1922 and 1924), an initiative of the Road Federation, went directly to the roads. This process also helped to redefine the car toward a socially accepted technology. In Sweden, as opposed to Norway and many other European countries, the process of domesticating the automobile was nearly completed around 1950.

By the WWII, the SRF had gained the position of a politically neutral provider of technical knowledge and expertise. The Federation was deeply embedded in the personal and professional network of the state's road administration and had become a respected and influential actor. It was reorganized in 1947 with economic support from chiefly the car and oil industries and the Federation's role as an umbrella organization for the wider car lobby was thereby reinforced. It was in fact its reputation as an apolitical mediator of technology that afforded the Federation its main resource as well as the principal reason for the participation of the other commercial members. After the war, the central interests of the car industry crystallized around a common strategy based on maintaining a united front publicly. The 'sound development' of mass motorization was the objective the industry could gather around. Lobbyists felt that the wisest path was to tone down internal divisions, for instance between commercial and private traffic, so as not to damage their common interest. And despite its apparent character as a commercial lobby group, the strategy of *technifying* the road question paid off once again and served as a sign of legitimacy which enabled SRF to initiate and heavily influence *Road Plan for Sweden* in 1958.

The strategy of *technifying* the debate on roads and streets had two advantages. First, the controversial character of the automobile was avoided by reference to allegedly neutral technical and scientific arguments. Second, by connecting roads and streets to a future of modern industry and infrastructural systems and the development of the well-fare state, public engagement in the sector was secured. In short, the articulation of publicness through the automobile was a ticket to success.

In a quite recent example (from an historian's point of view), the Swedish prime minister Tage Erlander corroborated this last point in a debate in 1956 with the right-wing leader Hjalmarsson discussing proposed state expenditures in the extensive national road plan mentioned above:

> Not even Mr. Hjalmarsson wants to go back and forth in his own courtyard, he wants to go out with his car on the roads. As citizens acquire a car, the demands on society's efforts are increased. (Blomkvist, 2001)

Having said this about the influence of the roads and automobile lobby, it must again be noted that they promoted a popular project. A private automobile was a goal for most people and Sweden quickly became the most car dense country in Europe in the middle of the 1950s. Furthermore, the roads and car lobby were certainly not alone in their efforts to turn Sweden into a car society. Many other influential groups such as experts in traffic and town planning, representatives from a united industry and not the least, almost every politician, both on state and municipal levels, endorsed the vision of a motorized future (Lundin, 2008; Wikman, 2019).

Turning to city streets, the technologies and tools used were similar as in road keeping, but originally the material used for street construction was logs of wood and to build streets was called 'to bridge' (broa) which is why street and road keeping in older laws often are referred to as 'bridge building.' However, already in the fourteenth century, cobble stones (kullersten, fältsten) were being used, at least in larger towns and in the most important streets. These streets had a width of 4.8 m while smaller alleys had a width of 1.8 m according to the City Law of Magnus Eriksson (1350). During the sixteenth century, stone pawing became more common and stone lined gutters were requested by the authorities. However, it was not until the middle of the nineteenth century that the use of cut granite (huggen sten) was established as the norm for streets with high traffic intensity. At that time, the innovation of special areas, footways, or sidewalks (trottoarer) finally were realized after many years of debate. Before this time, pedestrians in the streets had to share the space with horses, carriages, and carts, which of course made a stroll through the city quite adventurous and nothing at all like a walk in the park.

Street keeping was also professionalized and industrialized just as in public roads. From the beginning of the twentieth century, the most important system builder for city streets was the Swedish Association of Municipal Engineers, founded in 1902 by high-ranking engineers in municipal administrations in Stockholm, Gothenburg and Gävle. Very soon other municipal engineers from Sweden's larger towns joined the ranks. Membership grew from 97 members in 1902, to 330 in 1925 and to 675 in 1950. The association was engaged in street building technology from the start, and they applied the same strategy of *technifying* the street question as the Swedish Road Federation (SFR) used in roads. One of the first issues was how street and curb stones could be manufactured. At the annual meeting in 1913 in Västerås, Malmö city's building manager raised the question on stones used for paving and footpaths and as edging and guttering. He noticed that both dimension and processing varied and if a Swedish standard could be established it would benefit all municipalities. A committee was appointed, and the results were reported 2 years later, at the annual meeting in Gävle. In 1932, yet a new proposal was presented in cooperation with the association's Danish, Finnish, and Norwegian equivalents. In 1961, the standardization work was moved to the Swedish Standardization Commission (SIS). Also other paving materials, such as asphalt, were discussed and in 1924 the issue of street and road paving economics was introduced. The purpose was partly to reduce maintenance costs and partly to increase traffic capacity to meet the expected boost in automobile transportation. These questions were closely related to the SRF's efforts in traffic calculating and prognosis. The municipal engineers presented their own car traffic calculations for Swedish towns in 1930. A handbook for street building and maintenance for the new car society was published in 1953 and in 1969. It basically covered everything from geotechnics to signal facilities and community planning. Another influential handbook published by the association in 1973 was 'Guidelines for the geometric design of streets' (RIGU 73) in collaboration with the National Road Administration, the Swedish Association of Municipalities, and the National Planning Agency. The content described how streets, intersections, cycle paths, and so on should be designed to cope with traffic loads and safety requirements (Tjulin, 2002).

Although street building and maintenance were at the center for the association, a variety of traffic issues were also important. Traffic noise, traffic flows, parking facilities, traffic rules, signs, and signals were areas covered. The municipal engineers often worked in close cooperation with members of the automobile lobby such as the RAC. The general direction of the efforts was how the streets and public places, and the whole town should adapt to different traffic needs, especially to the private automobile. In 1960, a whole conference week was devoted to the car question. The issues centered on the building of the car society; what could be done and who should pay for its realization. Many speakers at the conference were from organizations within the car lobby and the municipal engineers thus showed that they were at the forefront of motoring and the construction of cities around the car. Also, in 1967, when Sweden switched to right-hand driving, the association was deeply involved. For example, they took the lead in the right-hand traffic investigation in Stockholm where 25 interchanges and more than 200 street intersections had to be changed, 176 traffic signal facilities were moved, and 35,000 road markings were repainted. The process of switching from left-hand to right-hand traffic deserves its own book. It included an enormous rebuilding of the whole road and street network and a massive national campaign to educate the Swedish people on the new traffic rules. In many ways, it was an effort like a national mobilization during times of war and it certainly was a climax in the articulation of publicness in the sector with the automobile as the obvious point of departure.

4.2 SYSTEMS CULTURE

In research on the history of infrastructure and large technical systems, there is an idea that a special *system culture* develops within each system. According to Kaijser (1994), this culture among the dominant actors is characterized by uniform education, common and overlapping career paths, and common views on what is right or wrong. The systems culture fosters a certain 'system rationality' or 'inner logic' which guides the thoughts and leads the actors to see solutions in line with the system rationality and logic. The system culture thus contributes to the system's inertia and creates path dependence. We can discover such a systems culture or system builder culture, both in roads and streets and in water and sanitation.

In the road and street sector, after the demise in the 1920s of the institutional conservatism discussed earlier, this culture grew strong and was shared by both state and private actors. Common educational background and cross-border career paths were factors that created a sense of belonging within the 'guild,' regardless of whether one worked in the commercial or state and municipal side of the road system. It must be emphasized that the system builder culture, even though it is community-creating, does not exclude fierce conflicts. System builder culture should be seen as a collection of informal rules for how conflicts are expressed, which opinions are considered grounded in science and facts and which are defined as not. It maintains an 'inside' and an 'outside' within the road system (Blomkvist, 2001). The sociologist of science Bruno Latour puts it this way about the 'exclusive network' of technicians and scientists and their privilege to specify what is rational and what is not:

> In this way, a scientific/technical discourse is created within the network that defines what is knowledge and what is belief, while at the same time an asymmetry arises between those inside the network and those outside … (which can lead to) … that the critics are either redefined as irrational and representatives of extreme views or they are convinced to accept the network's discourse and thus become allies. (Latour, 1987)

The most important components of the road system building culture have been a technical, scientific view of the field of road traffic and a common value base created by a similar educational background. Both were based on the professionalization project of the road and municipal engineers. As been discussed, the *technification* of the road system was intimately connected with the road engineers' desire to expand their professional field. This applies both to the 'industrialization' of the roads in the 1920s and the full-scale adjustment to mass motorization in the post war years. The Swedish Road Federation (SFR) and the Association of Municipal Engineers became the most important organizations for this endeavor, and in SFR, its commercial members thereby gained access to an important political resource: the reputation of engineering science as politically neutral.

As already discussed, after WWII, the system builders embraced the idea of a car society, which Sweden became from the early 1950s. However, it was apparent that to implement the desired car society, Swedish planners and road engineers had to draw upon knowledge from outside the nation. The inspiration and knowledge to adjust the Swedish motor road system to meet the demands posted by mass motorization came from 'American Traffic Engineering' which at the time was a new way to handle traffic. The goal of traffic engineering was to create free and unrestricted traffic flows and to meet the demands of mass motorization by increasing road capacity. Its core message was the often-quoted phrase: 'All traffic demands should be met.' Traffic engineering can be said to be the transport sector equivalent of Scientific Management or Taylorism and it became the foundation in the so-called 'predict and provide ideology.' Traffic engineering identified the critical problem in the motor road system as lack of flow and gave the solution: expanded road capacity. The carriers of this new car orientated culture were the interest groups mentioned above, SRF and the municipal engineers, commercial interests and state and municipal authorities. Knowledge in planning for the car society, adding to the systems culture, also came from other professional fields such as traffic and city planning experts as well as the field of Human geography (Lundin, 2008; Wikman, 2019). It is fair to say that all important actors, including politicians, embraced this invigorated automobile-based road and street systems culture. From now on, the car society was firmly established in the articulation of publicness in the road and street sector (Blomkvist, 2001, 2004).

doi: 10.2166/9781789063981_0041

Chapter 5

Pre-modern water and sanitation

In this chapter, I discuss and analyze the pre-modern history of water and sanitation (WS). The field is large, and I can only touch on the most important aspects. I briefly relate the ancient legacy before turning to pre-modern WS. I discuss to what extent the pre-modern history has affected modern service arrangements in Sweden, and my focus is on the question of whether water provision and sanitation has been articulated as a public or private responsibility.

One interesting area outside my scope is the contested theories by historian Karl August Wittfogel who coined the term *hydraulic civilizations* for early state power residing on the control of water resources like advanced irrigation schemes. Even though drinking water is not in focus and the term does not fit in pre-modern or modern European history, the impressive water systems discussed by Wittfogel surely point to an interesting trait in modern WS: the fancy for large-scale, centralized and piped solutions (Wittfogel, 1957).

5.1 PRE-MODERN DRINKING WATER

The most famous constructions for water provision are the Roman aqueducts, although there are older examples of impressive water systems. In 33 BC, over 500,000 cubic meters per day were transported into Rome by the aqueducts. Water was mainly collected from wells and springs and not from surface water like rivers as one might believe. Furthermore, contrary to common imagination, most aqueducts were not elevated structures but ground level canals or underground tunnels. Water was led in a gravity system according to the flow-through principle, without the possibility to turn off the flow. With the great amount of water coming in, the diversion and drainage needs were of course huge (see below on Cloaca Maxima). Around 60% went to public facilities like fountains, 20% to the emperor's palace and 20% to private homes. What is clear is that only wealthy families in Rome had running water and private baths. The famous fountains like Fontana di Trevi, still in use today, had a double function. They provided water for the populace, and they were also leveling reservoirs or safety-valves to even out water pressure. Each of the aqueducts ended with a large final fountain and smaller fountains were scattered along the water course of the aqueduct. Some of the ancient fountains are still part of Rome's water and sanitation system.

However, the aqueducts were not only providing drinking water. The copious amount of water transported by the aqueducts were mainly used for the Roman baths:

> 'At all periods of Antiquity, wells were a prime, if not *the* prime resource. Aqueducts, which we often think of as an essential feature of any ancient city – certainly of any Roman one – were often very late in arriving on the scene, for it was usually the opening of large bath complexes, voracious consumers of water, that spurred their construction. Previously, all water supplies had to come from cisterns and wells, and even after the advent of the aqueduct would remain in service. Indeed, in most cities it is a good question whether one should think of domestic wells as supplementing the aqueducts or the other way around. And of course some cities, such as Ampurias and London, never had an aqueduct at all, and relied on cisterns and well-water all through their history ... (when an aqueduct was built) ... some of its water might be diverted for drinking and domestic use, (but) its major purpose was normally to supply the baths. Seen in this light, the Roman aqueducts become largely a social and recreational facility ...' (Hodge, 2000).

Nevertheless, it is important to note that Roman law had many regulations on water issues, including drinking water, and Rome also had a quite impressive administrative apparatus managing water provision. As in the Roman road administration, the highest-ranking water officer was the curator, *Curator Aquarium*, who oversaw the upkeep of the aqueducts. A central task for the curator was to guarantee the flow of water day and night and to manage the constant need for maintenance (Deming, 2020). The curator also had an extensive workforce at hand which included occupational groups specialized in various parts of the Roman water system (Crow, 2012; Mays, 2010).

But despite all this and although Rome had left us with an impressive heritage of water provision, the general view among researchers is that although famous and celebrated, '... it is not correct to talk about a technical or institutional trajectory stretching from Rome to the early nineteenth century, the arc was broken for 500 hundred years or so' (Goubert, 1989).

However, references to Roman water and sewage technology were often used in the rhetorics of advocates of gridded systems such as Wilhelm Lejonancker who designed the piped water system in Stockholm (see below). French historian Goubert (1989) notes that this admiration for Rome was based on a superficial (smattering) knowledge of the Roman water system: 'With their smattering of Greek and Latin culture, mayors, members of parliament, doctors, architects, and engineers, whether in France, Western Europe, and North America, were often inspired by the Roman model.'

Having said the above, it must be noted that references to Rom was not empty rhetoric. The role model borrowed from Roman water systems played a substantial role in, for example, the construction of the Paris sewage system in the nineteenth century. Its rhetorical or symbolic meaning was probably instrumental in mobilizing actors and capital (Reid, 1991).

Although Roman technology and organization lay dormant until European state and municipal governments had the motivation and the resources to do anything about water provision and sanitation, a substantial institutional and legal legacy survived, in the form of Roman water legislation. The laws in Europe which regard water as a public resource with ownership only of its products (usufruct) derive from the Roman tradition, while those emphasizing land ownership bordering to the water curse as the guiding principle derive from British common law. During colonialism, the European, and thus partly Roman, water law was exported to all corners of the earth (Dellapenna & Gupta, 2009). Nevertheless, it is my impression that these Roman laws mainly targeted water issues other than those related to drinking water. As will be discussed later, early water legislation in Sweden did not mention drinking water at all but instead were directed toward so-called *defensive* (land drainage) and *lucrative* (waterpower) projects and focused on ownership and the right to manage water resources for farming and industry. It seems to me that literature on the history of water legislation often blur the lines between drinking water and other legal areas where rules on water were included. One important legal legacy concerning drinking water that has survived though is the right to draw a water pipe over land owned by another. These rules have been a fundament in drinking water (and sewage) legislation since at least Roman times where it was called 'Aquae ductus' (the right in law to carry water by means of pipes or conduits over or through the estate of another) (Blomkvist *et al.*, 2023).

In the period following the fall of the empire, Europe saw few attempts to systematically tackle water provision and the achievements of Rome were largely forgotten. However, Moslem Spain was unusual with its twelfth century system in Seville which is considered as a remarkably complex system for the collection and inhouse-distribution of water. But most European towns relied on springs and wells for drinking water. Concerning water quality and pollution many cities, including in Sweden, tried to prevent dumping of refuse and sewage in rivers and lakes through legislation but technical or systemic solutions were rare. As discussed, the state and the municipal authorities had no power to intervene in what was seen as private matters. According to the ideology of what later would be coined *night-watchman state*, authorities did not meddle; the welfare state was still a long way off and it is up to private individuals to cater for their own (Goubert, 1989).

In Sweden, the situation concerning provision of drinking water was much the same as in the rest of pre-modern Europe and very different from water and sanitation in ancient Rome. As touched upon, in the countryside and in the villages, many public matters such as roads, land drainage, grazing, fishing, hunting, and common forestry were managed as Common Pool Resources (CPRs) and regulated in the village by-laws or village ordinances (Byordningar). However, drinking water was not mentioned at all (the same goes for sanitation). I have investigated approximately 400 village by-laws published by Ehn (1982) and searched the web for other village ordinances and found no sign of drinking water being a public or common issue.

The village by-laws were codified and standardized in the so-called model village by-law in 1742 and included the most important stipulations needed for the village council and the village eldest to manage CPRs (Blomkvist & Larsson, 2013; Ostrom, 1990). The purpose of the model village by-law was to make sure that conflicts within the village around these types of common resources were solved within the village community and not referred to the County court. Water issues were only mentioned in Sections 5–7 where drainage of land using ditches were mentioned, in Section 26 public water for fire protection was dealt with, in Section 33 regulations for public management of wells, springs or 'other common water reservoirs' were listed, but, and this is important, only for cattle, as shown in this example:

> 'All villages should be well supplied with water, and to that end all wells, springs, or other common watering-places for the *cattle*, shall be constantly kept under control, and when it is necessary that they should be cleaned or dug up, no one may escape from them at (?) silver coin's fine, if someone does it, the village eldest and the village council have the power to let the reluctant hire workers then collect the wages and fines from him.'

But Section 33 was included in just a few of the village by-laws related by Ehn and similar wordings on water for cattle can only be found in around 5–10% of the approximately 400 individual village ordinances. To conclude, there is nothing in the older history of rural areas, such as village ordinances or by-laws, legislation, or court proceedings, that treats drinking water as a public concern. Drinking water provision in the countryside of Sweden was simply not regarded as a CPR in the meaning of Ostrom (1990), and it was not articulated as a public concern, at least not in any formal sense in written laws and regulations. These findings are surprising and challenge a belief that drinking water provision was a communal undertaking and that publicness in water was articulated very early in Swedish history. I think this belief is quite common, at least it was mine before this project, and it is probably based on the fact that so many other areas in the villages were managed communally.

I cannot really say if the situation in Sweden was unique because I have not made any systematic attempt to find out if pre-modern drinking water provision has been regarded a CPR internationally. Shiva (2003) writes interestingly about water as a CPR in India. However, this applies to water for irrigation. No CPR organizations specifically for drinking water (or sewage) existed. However, if there were a CPR-managed irrigation system, drinking water was also included as a side effect, but no specific arrangements were made to handle this part of the resource. In a very interesting investigation on household water provision in the countryside in Sakumaland, Tanzania, Drangert (1993) shows that even if drinking water/household water was considered a CPR, it did not mean that it was actually managed as a CPR: 'Household water remains a common pool resource which means that everyone is entitled to draw household water from any water source and that it is monitored by all residents in the area.' Even if this norm was supported by most, many other factors, such as fear of 'free riding,' often prevented communal or cooperative solutions from becoming a reality. Drangert concludes by stating that these norms could well be general: 'The pronounced Sukuma norm that water is a CPR from which to draw household water is believed to be the general pattern in rural areas in most of Africa.' In contrast, interesting case studies of elaborated and advanced CPR-management of both drinking water and water for irrigation can be found in Yemen dating back to the fifteenth century and in Tashkent, Uzbekistan, from the eighteenth century (Hehmeyer, 2007). Another interesting case comes from Mexico where Indigenous people in the Meseta region since a very long time, at least before the sixteenth century, has managed to cope with harsh conditions and limited amounts of water by developing sociocultural strategies for water management based on the following three components (García, 2007):

- The emergence of what we may call a 'culture of water scarcity' utilizes modest volumes of water due to the lack of adequate sources of supply.
- A form of social organization permits 'community control of water,' where this resource is seen as a collective good to which the entire population must be assured access. In addition, all members of the community share the responsibility of conserving and maintaining the sources of supply and for capturing, transporting, and distributing water.

The emergence of a 'culture of ecological water use and management' was associated with the people's cosmovision (worldview), where water is highly valued and must be cared for because it is a 'fruit' bestowed by 'mother nature.' This attitude is reflected in the practices of water use and management, whose basic ecological principles are low consumption patterns (little waste), the diversification of sources of supply (utilization of rainwater, springs, watering holes), multiple applications (productive and domestic uses) and recycling (minimal discharge).

The best explanation for the difference between Sweden and the cases related above is water scarcity. In regions where water has been a scarce resource, it is possible to find elaborate schemes for communal, CPR-like, drinking water management. In most locations in Sweden, water was not a scarce resource and there was no need for cooperation and public or common management. This notion is corroborated by Drangert (1993): 'Norms about water rights do not make themselves manifest as long as water is abundant, but their sustainability is put to the test when water gets scarce.'

Despite the statements above and although it is quite clear that drinking water was not a formal public concern in the Swedish countryside, this picture is nuanced by evidence from research using ethnographic and oral history sources. Even though formal rules were missing in the village by-laws, there seems to have been informal rules concerning water as a CPR. Many informants testify that it was seen as rude and breaking social norms to deny others to use water from a private well or spring. People in the villages were expected to share (Drangert, 1991). Interestingly, these social norms of water sharing seem to be universal and historically stable through different cultures. As mentioned, almost the same widely accepted expectations were present in, for example, traditional rural communities in Africa where you had to share the water with other people, but not with other people's livestock.

In towns however, due to higher population density, the situation was somewhat different. Through the seventeenth and eighteenth centuries in Europe, water provision in towns sometimes relied on *proto systems*, which was rudimentary technological solutions using water pipes made of wood (drilled logs) (Hallström & Melosi, 2022). Thus, in some Swedish towns, at least in the larger ones, attempts were made using various types of proto systems for piped water during the Middle Ages and the following centuries (Uppsala, 1640s; Gothenburg, 1780s). In Malmö, an impressive stretch of wooden water pipes has been found, dating back to around the early sixteenth century (Person, 1999). In Stockholm, in the beginning of the seventeenth century, a few attempts were made by private citizens to construct small water distribution systems by wooden pipes and to build water pipes supplying the Royal castle and other important buildings. However, these attempts were not successful (Cronström, 1986). In Uppsala, in 1642, a water main with iron pipes was built, supposedly the first in Sweden which led water to the castle. Due to lack of interest on the part of the residents, the pipes were forgotten. The naval town Landskrona, on the east coast, had a rudimentary water main during the late seventeenth century which was replaced when the first modern water system was built in 1869–1874.

The most famous and well researched proto system is the so-called Kallebäck pipe (*Kallebäcksledningen*) in Gothenburg (Bjur, 1988). It is interesting because it shows an attempt to provide water for all citizens in a town before the municipal authorities really had the ability or will to provide public drinking water in our present-day understanding of the concept. The spring that became the water source is mentioned for the first time in 1692, and in 1714 came the first proposal to transport the spring's water in pipes to the town. However, this project never materialized. From the middle of the eighteenth century, private water peddlers delivered the spring water and sold it in the city. In 1785, a group of the cities wealthier residents turned directly to the King for permission to build a water main, which was granted in a Royal letter. A water directorate was formed and given the right to expropriate the land required to lay out the water main because landowners at the stretch of the pipeline refused to give up their proprietorship. Bjur (1988) makes a strong point of this expropriation right. In the Royal resolution, the water main was seen as a public good (Commune bonum) like public roads and therefore the right of the town should override the right of the property owners. Bjur's hypothesis is that the Royal decree can be seen as a forerunner to the so-called 'line right' (ledningsrätt) which is the present-day legal foundation for municipalities to build water pipes on private land (like the Roman 'Aquae ductus' and the 'right of way' in roads). The financing of the project came partly through grants from the city with a third, and the rest through gifts from the city's more affluent residents. In November 1787, the king Gustav III inaugurated the water pipeline. It was built of hollowed and buried logs of aspen, pine, and oak and stretched for 4.8 km to a water cistern with a fountain at the city center where citizens could collect their drinking water. Quite soon though, the project ran out of money and after some initiatives to privately finance the operations, the king decided that the water pipe should be paid by taxation in 1804. However, the obligation to pay tax in cities in this time, before the Municipal reform of 1862, was not laid on all citizens but rested on the Burghers and the property owners.

The Kallebäck pipe is surely an interesting and quite unusual example of public involvement in the history of piped water. The ambitions of wealthy people and city authorities in Gothenburg to provide for piped water and the expropriation rights given by the Crown point to legislation that came to be enforced in both roads and streets and water and sanitation some hundred years later, and surely shows an early articulation of publicness. It is not possible for me to ascertain if the Gothenburg case really is a forerunner to public water provision or if it is an isolated example. I recommend further research. Still, I believe it is fair to say that this proto system can be seen as something in between older forms of service arrangements in water provision and modern infrastructure. It is perhaps the *missing link* in the evolution of municipal infrastructure. The private initiative for piped water in Gothenburg is like the early development of a 'lighthouse system' in Sweden during the eighteenth and nineteenth centuries. A coastal lighthouse is per definition a public good as its services can be used by all seafarers in need of guidance (not excludable, non-rivalrous). The Swedish state was reluctant to pay for lighthouses and it was merchants in the coastal towns that often took the initiative to secure their trade by ships at sea. Lighthouse development shows an evolution of mixed institutional design between private production of a public good and government involvement. Lindberg (2015) discuss the importance of 'civic solidarity' and 'bourgeois virtues' when merchants voluntarily financed lighthouses. The same civic solidarity and bourgeois virtues were displayed in the building and financing of the Kallebäck pipe. Perhaps one can say that these proto systems, financed and managed by private individuals of some wealth in cooperation with city authorities, were early examples of what we today call public private partnerships. However, I believe that it is not correct to describe them as public infrasystems anywhere near our modern meaning. The water delivered by these early piped proto systems were certainly for all to use but the service arrangement was still not fully publicly financed and managed. Regarding the discussion on *publicness*, it is important to note that the proto systems discussed above, and the communal wells described below, were public in a weaker sense than the modern water systems being realized later. Hallström (2003) summarizes the history of drinking water like this: 'Before the 1860s, the primary ways of obtaining fresh water in urban areas was from wells, springs, and waterways. There were a few smaller water pipes in certain cities, the most famous one being *Kallebäcksledningen* in Gothenburg, but modern piped systems were wholly missing, and urban residents had mainly to resort to manual, decentralized water supply.'

Water peddlers were probably common in Swedish towns although I have only found few examples mentioned, most often in passing, in the literature on water history. One 'water delivery man with his horse and barrel' (Vattuköraren) is mentioned in a famous song by Carl Michael Bellman (1740–1795), *Fredman's epistle no. 48: Ulla Winblad's journey home from Hessingen in Mälaren one summer morning in 1769* (verse 18). Water delivery men were also enrolled in the firefighting brigades in Stockholm, and as mentioned, water was sold in Gothenburg from the spring that later became the source for *Kallebäcksledningen.* Water stores serving bottled and sometimes sparkling water were set up from the middle of the nineteenth century. One example was the water store in Kungsträdgården, in the center of Stockholm. The shop opened in 1850 and closed in 1933. In this very store, a literary murder takes place when the Reverend Gregorius is poisoned in the novel *Doctor Glas* by Hjalmar Söderberg (1905). It must be noted that the proto systems and the possibility to buy water from water peddlers, in stores or to use private wells, were a prerogative for the rich. Poor people had to rely on the water in springs, lakes, and common wells.

In Stockholm, Cronström (1986) presents some facts on common wells around 1600. They were spread over the town for fire defense reasons. Nevertheless, people were free to fetch water, although some evidence points to the fact that the water was 'hard' (rich in minerals) and not liked very much. In the middle of 1700, J. E. Carlberg, the town architect, listed around 20 wells and other water catchments that were to be used for fire protection. At the end of the seventeenth century, there were nearly 300 wells in Stockholm, most of them private. In addition to these, there were around 25 common wells in the city, which supplied the population with water and whose locations were decided upon with fire protection in mind. This number increased significantly during the nineteenth century and even in the 1860s, that is at the same time as the city's piped water supply system was built, new wells were dug or drilled. Thus, Stockholm had, during the entire period before the middle of the nineteenth century, several common wells, which in terms of drinking water, supplemented private wells and water collection from Lake Mälaren. In 1858, when the piped water system already had been decided on, there were approximately 50 common wells in Stockholm where private citizens could fetch drinking water. Furthermore, there are evidence that the authorities did take some responsibility even before the 1800s. In 1739, for example, in a Royal decree to the city, magistrates are stated that the so-called 'building and civil service college' were to oversee the quality of wells and drinking water (Dufwa, 1985). Another even older example can be found in Pettersson (2008) who relates a 'city council protocol' from 1640 referring to discussions on how to deal with provision of drinking water in Stockholm. According to the protocol, the council made a monetary contribution to build four additional common wells.

Some researchers interpret the attempts by the municipal authorities related above, as an indication that '… water supply was the first *necessity of life* (my italics) that became the subject of common solutions for "the public good" when already in the 18th century they started distributing well water to collection points in Stockholm through wooden pipes' (Bjur & Malbert, 1988; Svedinger, 1989). The proto systems and common wells for water supply were certainly existing but they were not designed solely to provide water as a *necessity of life.* Drinking water was not the only, or even the main, reason for building and maintaining proto systems and public wells. Rather, public drinking water could be seen as a welcome side effect of the basic motives for public water provision: fire protection and street cleaning. Thus, with a possible exception from the situation in Gothenburg, in the articulation of publicness concerning water in its many meanings and forms, fire security and tidiness were articulated more strongly than provisioning of drinking water for the people.

In conclusion, it is difficult to find evidence for a strong public ambition on behalf of city authorities to provide arrangements for drinking water before the beginning of the nineteenth century. There were attempts made but compared to street cleaning and fire protection, water provision was not a priority. Furthermore, the articulation of publicness in drinking water was much weaker than in other sectors such as roads and streets and even compared to sanitation, which will be discussed below. I have found no evidence for the existence of a serious obligation from state or city authorities to provide for drinking water and there was not really an expressed public desire to provide. Public drinking water was not mentioned in the medieval city-laws or in the law of 1734, although roads, streets, and sanitation (outer) were directly targeted as public concerns. Furthermore, water provision is not mentioned as a formal public obligation in research on the role of city magistrates and councils of Burghers, the century old authorities of the towns, in urban history literature or in contemporaneous reports and literature. I am aware that it is dangerous to draw conclusion based on silent sources, but in this case, it seems to me that the sound of silence is loud and clear. To sum up: in towns, the early and scattered attempts on proto systems were rarely successful in the longer run and common wells were not only for drinking water provision. It is hard to interpret their existence as signs of high public engagement in drinking water because publicness in drinking water was not at the political agenda and they certainly did not transcend into the modern piped system of the late nineteenth century.

That water issues in general and especially drinking water were not formally regulated in the written laws, but instead followed informal and older established customs; becomes clear in this historiography of Swedish water legislation by the Swedish Environmental Protection Agency (2008, my italics):

'Swedish water law is based on a private law system, that is that the control over the water in lakes and waterways is linked to the ownership of individual land. The perception of the right to water according to private legislation, is very old in Sweden. It is already clearly expressed in *Hälsingelagen* from the 14th century. There you find the well-known regulation that "the one that owns land also owns water". The regulation of the Hälsingelagen was also supplemented with a provision which allowed the beach owner to sell his right in the water. A similar view can be traced in the *Västgötalagen* from the latter part of the 13th century and is reflected in the national law of 1734. The concept of availability is the same for groundwater. The availability right does not prevent the water, *according to old customary law, being used by others for washing, fetching water, watering livestock and the like.*'

It is my belief, although it is not explicitly mentioned in the sources or the literature, that drinking water, in both towns and in the countryside, was regulated by informal rules based on old customs. Perhaps drinking water was mainly seen as *foodstuff* and thus per definition belonging to the private sphere. Public authorities did not engage in the food habits of the people. This notion that water was food is strengthened by Drangert (1993) in his investigation on water norms in present-day Tanzania (with reservations for geographical and historical distances): 'Water is part of the diet, but the informants treated the provision of water separately and differently from other food items The Wasukuma do not expect the district council to provide food unless there is a serious famine.' Today drinking water is certainly seen as food and some regulations therefore fall under the Food Act (SFS 1971:511). Since the act does not apply to the handling of food in individual households, it does not apply to water from individual wells (Christensen, 2015). And as mentioned, another possible explanation for this surprising finding, at least from a modern perspective, is that water in Sweden was considered and everlasting resource. Drinking water could be found almost everywhere and was far from scarce. Finally, and to end this section, I want to highlight that my conclusions on a relative low level of public engagement in drinking water provision are based mainly on secondary sources and refer to larger towns. It would certainly be very interesting to test this thesis on primary sources from various municipal archives around Sweden. I strongly encourage further research along these lines.

5.2 PRE-MODERN SANITATION

To understand the modern water and sanitation system based on water and sewerage in underground pipes, it is important to note the varying meaning of the term *sanitation* in different historical periods. First, we must separate between *inner* and *outer* sanitation. Inner sanitation was the handing of human refuse, such as dirty excess water, household leftovers, and human and livestock excrement. These practices were for a very long time seen as an exclusively private matter. Inner sanitation and excrement handling relied on latrine pits or cesspools and were dealt with inside the confinement of the private property. According to Wetterberg and Axelsson (1995), hygiene, cleanliness, and handling of feces was a private matter right up until the sanitary revolution from the middle of the nineteenth century. This applies to both rural and urban areas. Concerning outer sanitation, the basic rule in cities and more densely built-up areas was that property owners were required to 'keep their front door clean.' These rules can be derived via the so-called Björkörätten and as far back as in Birka of the Viking Age. In Magnus Eriksson's city law (1350s) and in the cities' own regulations ('Burspråk'), these rules were repeated. Outer sanitation included street cleaning and storm water management (and solid waste management which will be left out). Outer sanitation has always been a more distinct public concern and national and city authorities have since the Middle Ages worked hard to make people take care of these matters. Sanitation in general has been a public concern in a higher degree than water provision because of its obvious *negative externalities* affecting others. Your neighbor suffers if you don't keep your property clean. Negative externalities were, and are, of course more prominent when it comes to inner sanitation and especially the handling of feces. Human latrine is not any type of dirt. As mentioned, latrine management has been, and I believe still is, associated with strong taboos and the desire to dispose of these substances as far away as possible has been unchanging throughout history. However, this does not mean that feces had no value. As will be discussed, latrine has, in periods, been used as a fertilizer in agriculture and today we can see a return to these ideas of material reuse and circularity in the sanitation sector.

In the following, I focus on sanitation defined first as street cleaning and storm water management using open sewers and gutters and second on the establishment of underground piped sewers for excrement management. I will also briefly touch upon the period when excrement was managed by bucket collection. However, there is a problem with the delimitation of outer sanitation as street cleaning and storm water management. Both areas are of course also connected to street management in a general sense. Street (and road) keeping involves building and maintenance. The latter has two parts namely *technical* maintenance meaning keeping the street in a functional condition such as repairing pavements and *general* maintenance such as street cleaning and snow removal. This distinction is valid today also. Municipal street management, building and technical maintenance, is regulated in the Planning and Building Act (PBL), but not street cleaning which is handled in the Street and Cleaning Act which gives the municipality responsibility for street cleaning in public places. It was preceded by the order and cleanliness legislation, including the General Order Statute from 1956. In the Environmental Code from 1999, the legislation was moved to the Act with special regulations on street maintenance and signage (1998:814), usually called the Street and Cleanliness Act (Tällberg, 2018).

On early sanitation, the most famous ancient example is again Roman, the Cloaca Maxima (600 BC), which was an underground tunnel used for stormwater drainage of the Forum Romanun. It was not primarily used for the handling of water borne excrement, even though some public privies were connected. After the fall of Rome, some of the Roman water and wastewater networks were used in southern Europe during the Early Middle Ages and old Greek and Persian systems were still functioning in Spain, Malta, and southern Italy. In northern Europe, ancient systems were operational in, for example, Paris, London and in Visby, Sweden. Visby, with its peripheral position in Europe, is interesting as it shows that advanced ancient technology for water cisterns, cesspools, and some rudimentary sewers for excrement were transferred and used far up north (Westholm, 1995). However, just as in early water provision, I agree with Hallström and Melosi (2022) in their conclusion that in general, Europeans (and Americans) had to abide on off-grid arrangements for sanitation during the Middle Ages and well into the pre-modern and modern eras.

The period after the fall of Rome was pretty much a long stretch of stagnation. Sewers before the mid-nineteenth century were generally built for land drainage and stormwater and we would not call them 'systems' in any modern sense of the word, and to conclude I turn once again to Hallström and Melosi (2022):

> The premodern attitude about the disposal of wastewater was thus mostly to drain liquid waste from wherever it was deemed unnecessary, which meant manual discharge in a street, cesspool, ditch, or pipe to the nearest watercourse. Before the mid-19th century, the sewerage of ancient civilizations, notably Rome, remained unmatched. The sewers of the early modern period were street gutters or drains rather than real underground sewers and the discharge of waste in the sewers was forbidden, although a great deal ended up there anyway.

In France, for example, some villages had common water supply, but drainage and sewers were mostly unknown. From around 1860, cooperation public on water supply was increasing for both cattle and humans. These installations were typically a fountain or a communal well, maintained in kind by the villagers. But drainage and sewers systems were basically unknown. The removal, transport, and disposal of feces only existed in larger market towns and the main cities. Leak-proof tanks and cesspits designed for waste and excrement 'were few and far between' (Goubert, 1989).

However, even if sanitation technology was at a level we hardly would accept today, as mentioned some parts of sanitation were still a public concern at least from the Middle Ages, that is outer sanitation such as storm water management and street cleaning. In line with the prevailing ideology of *Politi*, state and city authorities tried, but rarely succeeded, to uphold order, tidiness, and cleanliness in towns. It must be noted that the regulations on public sanitation were only issued for the cities. In the countryside, village by-laws and other sources are silent on outer and inner sanitation.

In Stockholm, and in other larger towns, the state and the city authorities tried to organize outer sanitation, but they were not very successful. In the Middle Ages, the building code of Magnus Erikssons city law stated that rules for the towns are to be given in the Burghers written ordinances for the city which included regulations about the cleanliness of streets, bridges, and yards. On latrine management, nothing else was stated than rules for the building of privies in Section 2 of the law: 'No one shall build a secret house next to his neighbor or on a public street, unless he leaves open three feet of drip space between his house and his own yard or house.' According to the customs of the Middle Ages, it was considered completely out of the question that an ordinary citizen would manage latrine disposal (Blomkvist *et al.*, 2023).

The authorities did not manage to implement efficient outer sanitation for many hundred years. For example, in 1711, when the plague ravaged Stockholm, several regulations on improved street cleaning were issued. According to Dufwa and Pehrson (1989), the fear of the plague seems to have resulted in only a temporary improvement. In 1723, city authorities launched a new plan to organize outer sanitation motivated like this:

> 'Since human health is generally maintained most of all by the purity of the air and water, the squares and streets, together with the bridges and

harbors, must be kept in proper condition, the impurity must be removed early in the morning from the road and nothing from the windows, whereby the walls are greatly stained.'

The organization envisioned concerned not only cleanliness but also the function of the police in various branches of activity, the supervision of trade, crafts, servants, poor relief, and the fire service, paving the streets and much else. The proposal aroused violent opposition from the Burghers, which would bear the costs and was never realized. The Building Code in the law of 1734 contained quite detailed provisions on public engagement but the legislators apparently considered that these problems were so complicated and varied in the cities that they could not be solved generally in the law. Instead it was left to each town to decide on these matters, including sanitation. It was not until in the Building Act of 1736 (see below) that these matters were regulated in national legislation.

In the eighteenth and early nineteenth centuries, Stockholm had a bad reputation for its dirty streets and smelly waterways. Already Carl von Linné (1707–1778) complained that he was sickened by the smell of cesspools and privies during the summer season. The wretched sanitary conditions were described vividly in 1815 by Johan Olof Wallin, the priest of Adolf Fredrik's parish, when speaking of the many waterways that received the city's waste. 'Day by day, year by year, the masses grow which, united by street filth, garbage, rubbish, and all kinds of impurity, will turn these lakes into hideous, swamps' (Bäckman, 1984).

To sum up: external, outer sanitation and street cleanliness had early on been regulated in various building and order statutes. But until the middle of the nineteenth century, the internal cleanliness in Swedish towns, the household waste and the latrine, was essentially an individual matter and the city did not intervene. The disposal of internal refuse was arranged with latrines and waste pits within the private properties, which were periodically covered over and replaced with new ones. Waste was also used as fertilizer in gardens in the city or dumped in or near water. Since the eighteenth century, many Swedish towns, with the help of contractors, had made sporadic attempts to systematically use the waste as fertilizer, including by collecting it in special storage areas in the city and then taking it out of the city on carts and barges (Sjöstrand, 2014).

doi: 10.2166/9781789063981_0055

Chapter 6

Specific contextual factors in modern water and sanitation

Adding to the already mentioned *general contextual* factors discussed above, in this section, I discuss some of the most important *specific* contextual factors that affected water and sanitation (WS). The importance of these factors grew over time and mainly influenced the development in modern WS from the beginning of the nineteenth century and onwards. The specific contextual factors are connected to each other and it is fair to say that they all lead up to the Health Act of 1847, which can be seen as the final articulation of publicness in the sector. The Health Act operationalized all the preceding contextual factors and suggested legislative and institutional solutions to the problems under debate. Furthermore, in the chapter on water and sanitation in the twentieth century, I include two additional contextual factors affecting the articulation of publicness: environmental concerns due to water pollution and the growing focus on circularity and sustainability.

6.1 THE DIVIDE BETWEEN THE PRIVATE AND THE PUBLIC

As mentioned, the municipal reform of 1862 and its following statues developed alongside the emergence of a more pro-active state in the second half of the nineteenth century which gave the municipalities the obligation and the tools to intervene in the private sphere. In this new political landscape, the issue of 'intervening vs. not intervening' was replaced with the question: 'Intervene, but in what?' According to Kilander (1991), the result was a changed relationship between public and individual spheres of society, which now merged into a common sphere. The debates on public intervention starting from the second half of the nineteenth century follow a common pattern according to Nydahl and Harvard (2016) where certain actors start lifting a problematic societal issue:

> 'A debate is held where the central dividing line concerns whether this is a societal problem, or whether it concerns individual perspectives and interests. In cases where the debate culminates in the fact that it is

certainly a general problem, plans are raised for interventions, in the form of support or regulation. Interventions or proposals for such in various sectors … are consistently justified by the fact that the solution of the issue constitutes a public interest, which justifies that the freedom or decision-making rights of the actors concerned can be affected or restricted.'

Concerning water and sanitation, the Municipal Act of 1862 gave local government extended power to intervene in previously private areas. The borders between the public and the private domains were challenged. Matters such as inner sanitation that had been strictly private for hundreds of years became an issue for local government. Following Drangert *et al.* (2002), questions like how far the authorities could go in deciding on how to draw water pipes and gutters and other earlier responsibilities of the individual property owners came up on the agenda. Changes occurred in the division between the public and private spheres. In the 'private sphere,' the state or the municipality could act only in an advisory manner but not directly interfere. This divide becomes even more important in the debate on the passing of the Public Health Act in 1874 (see below). When the new act was designed, the committee crafting its text investigated the conflict between private property and communal collection of garbage and latrines and concluded that many house owners saw their properties as protected areas, where the authorities had no access. However, this situation was bound to change when people realized the necessity of giving the public the right to regulate and control cleanliness in towns and other densely populated areas. To limit public action only to outside areas was fruitless because the worst sources of sickness and *miasma* were most often found inside the house (Nelson & Rogers, 1994). Thus, in this new order, the distinction between the public and the private weakens and it became legitimate for state and municipal authorities to interfere.

6.2 THE SOCIAL ISSUE

The interest and the will to provide for public water and sanitation was motivated by a mix of both genuine concern for the living conditions of the lower classes and a true desire to improve their livelihoods and simultaneously by a fear of uprising, chaos, and revolution. It is easy to agree with Nelson and Rogers (1994) when they state that the work and the ambition to create the first comprehensive Swedish public health law in 1874 show that the authorities had a genuine concern for the working class and for the poor. Another example of concern can be found in the investigation of urban sanitation in Stockholm and Norrköping by Drangert and Hallström (2003). The health board in Norrköping complained about the large number of pigs in the city in a letter to the governor in June 1885 and argued that

> '… one could rather think that the three hundred people keeping pigs are violating the rights of the other twenty-seven thousand residents, because they do not enjoy the privilege of inhaling air, which is not been plagued … In a few words: the workers and their families may well enjoy the odor from the pig pens, but only exceptionally the aroma from the pork that is produced.'

The concern for, and the fear of, the poor and the working classes is often labeled 'the social issue' in historical accounts of the first half of the nineteenth century. The focus on the connection between a sound lifestyle and good health grew out of older views associated with the so-called 'patriarchal care.' The process can be described as a shift from the mill patron's responsibility for the employees to the municipality's responsibility for the entire city population. The social issue can also be interpreted as moralistic strategy. Laws and regulations became important in the upper class's project to control and direct the lower classes. Nevertheless, regardless of underlying motives, there is a consensus among researchers in that gradually, from the middle of the nineteenth century, responsibility for the welfare of the population was shifted from the patriarch to society. This change also meant that people in power, economically and politically, started to see a positive correlation between a healthy workforce and higher production capacity which in turn would affect the nation's general income and welfare (Wiell, 2018).

The fear of revolution and an uprise of the working class was a distinct feature in the motives behind the articulation of publicness in water and sanitation. Rosen (2015) describes these sentiments like this:

> '... fears of the revolutionary movements brewing in Europe in 1848, as much as the dread of cholera, prompted public health reforms. Each nation had intellectuals who pointed out the connections between ill health and poverty and demanded radical or revolutionary change as an answer to the problems of endemic and epidemic diseases. Friedrich Engels in England and Rudolf Virchow in Germany, for example, used public health as a focal point for demonstrating exploitation, dramatizing unhealthy social conditions, and demanding more democratic solutions.'

The revolutions of 1848 occurred in several European countries, starting in France in February. The wave spread all over Europe, more than 50 countries were affected and it is probably the largest wave of revolts in European history. Three main factors explain the revolutions: widespread dissatisfaction with the political leadership, demands for a higher degree of democracy and working-class discontent. Tens of thousands of people were killed and not much was won. From a Swedish perspective, the most important effects probably were the ending of absolute monarchy in Denmark and of the Capetian dynasty in France. In March 1848, Stockholm got a taste of the unrest. Anti-authority posters were put up demanding reforms and suffrage. The same day an angry crowd gathered in one of the town squares, protesting and smashing windows, and in the evening a crowd gathered around the Royal castle. The following days, riots continued, and the King finally gave the order to fire. In total, 18 people were killed and many wounded. At the same time as the riots in Stockholm, Milan, Munich, Copenhagen, and many other European cities were affected by similar violence and newspapers reported a revolutionary fire that could destroy the continent's old regimes. However, the revolutionary wave of 1840 was not the only reason behind the fear capturing the Bourgeois class. Already in the 1830s, Swedish observers noted signs of social crisis including high levels of poverty and criminality where large parts of population fell into unemployment, where household dissolved and morality was lost in filth and crudeness. In the revolutionary years of the late 1840s, these impressions grew stronger, and society seemed to balance on the brink of disaster. However, it is a bit paradoxical that the social question mostly centered around urban life, considering that Sweden, Stockholm being the only exception, really did not have any large urban centers at the time: 'Without any dramatic change in terms of population taking place, however, the cities and above all Stockholm during the period came to be perceived as a particularly vulnerable focal point of disorder and decay ...' (Björkman, 2020).

Nevertheless, it is true that the riots in Stockholm targeted injustices and society's elites during the 1830–1840s and sometimes carried clear political agendas. The lower classes rioted several times protesting against, for example, employers' wage-pressing tactics of calling in cheap labor from the countryside. Also the earlier July Revolution in France, 1830, and the Chartist movement in England fueled the fear of the working class. Many started to discuss negative sides of industrialization and began to picture images of large-scale acts of violence. These new fears of the workers were alloyed with century-old images of the lower classes as irrational, easily manipulated and explosive. Thus, the treat of the working class focused on Stockholm as a place of destructive activities portraying workers as socially and morally lacking, impulsive and disobedient, although without a clear political agenda or self-awareness. Workers were dangerous and at the same time described as 'faithful children' or 'inferior species' (Björkman, 2020). The social issue also had effects outside health and welfare connected to water and sanitation. During the second half of the century, several laws targeted the less well-off classes aimed to strengthen and facilitate, and also to prevent and control undesirable behavior. Examples of laws and regulations that were added in the wake of the social issue were the 1833 Lease charter, the 1842 Folk school charter, the 1846 Defenselessness statute, the Poor welfare ordinance from 1847, the abolition of house chastise in 1858 and the 1885 Vagrancy law (Wiell, 2018).

It is of course hard to tell if the fear of the poor and the lower classes was real or imagined. Nevertheless, it is quite clear that these emotions strongly influenced the motives behind the growing public interest in water and sanitation in Sweden. According to Hallström (2003), in his investigation of Norrköping and Linköping, the 'social issue' also rested on British experiences and rhetoric around sanitary reforms, where the fear of a working-class revolution was evident (more on this later).

6.3 HEALTH AND SICKNESS

From the beginning of the nineteenth century, the 'social issue' was blended and invigorated by concerns around high mortality rates, dramatic outbreaks of cholera and a gradually evolving new perception of the causes of health and sickness, which culminated in the so-called *Sanitary movement* originating in England around 1830.

Many died in Sweden, and they died young. Especially Swedish towns had a high level of mortality compared to European figures, despite their relatively small size. At mid-century, two-thirds of Swedish towns had crude death rates exceeding 23 per 1000, a rate used as the cut-off point for 'excessive' mortality in British urban areas. The committee preparing the Health Act of 1874 argued that Sweden's high mortality rate was due not to climate, physical geography, economic or social structure but of unsanitary conditions in towns. In a study of Swedish urban mortality, the Committee found that the situation in the towns was in fact worsening. During the period 1846–1851, mortality in the countryside was 19‰ and in towns 29‰. During the next five years, the mortality gap increased, 20‰ in the countryside and 33‰ in urban areas. There was a considerable amount of variation among Swedish towns: four cities had death rates exceeding 40‰, 15 between 30‰ and 40‰, 43 between 23‰ and 30‰, and only 25, mainly small towns, with crude death rates of less than 23‰. Stockholm topped the list with a crude death rate exceeding 45‰ and a life expectancy at birth of less than 17 years. This chocking number was an average value related to the fact that many children died before the age of five. If you survived these critical years, life expectancy of course got higher. With indications that urbanization was on the increase, the Committee clearly made its point. It summarized the result of its study:

> '... the comparison (with other European countries) is disheartening, yes even degrading, for Sweden cannot nor should not be denied. It is with sorrow that one must admit that at least part of the race, originally so strong and hardy, stands, as far as sanitary conditions are concerned, on the bottom ring among the civilized nations of the world' (Drangert *et al.*, 2002; Nelson & Rogers, 1994).

Cholera is often mentioned as an important cause for the establishment of WS systems in European cities during the nineteenth century. This terrible disease indeed played a part and often was the trigger for public interest. Cholera outbreaks hit Sweden 11 times between 1834 and 1874 and around 37,000 people died. The worst epidemics occurred in 1834, 1853 and 1866 and after each outbreak the debate on sanitary issues became fierce (Wiell, 2018). But the concrete fear of cholera is perhaps not the only explanation. Hamlin (2009) criticizes a simplified cause and effect relationship between cholera and public water and sanitation efforts. Hamlin claims that it is more fruitful to investigate the financial, political, and administrative foundations of these endeavors. He coins the term 'cholera forcing' to describe the idea that cholera outbreaks force beneficial changes in public health. Hamlin calls this the 'myth of the good epidemic' (Hamlin, 2009). Hamlin's thesis is quite radical and I cannot ascertain the possible existence of 'cholera forcing' in Swedish society. But it seems to me that the fear of cholera was a real and important factor in the articulation of publicness in infrasystem development in Sweden. Nevertheless, I focus on the other factors called for by Hamlin: '... the financial, political, and administrative foundations of these endeavors.'

Another fundamental factor that influenced the movement toward public WS was a new perception of health and sickness from the beginning of the nineteenth century and maturing in the 1830s: The *prophylactic* view was the insight that sickness could be prevented, and that good health, fresh water and sound sanitary conditions were tools preventing disease. Going back in time, historian Rosen (2015) relates the medieval Cristian view on sickness as the result of individual sinfulness: '... health problems were for the most part considered and dealt with in magical and religious terms ... At the same time, Christianity held that there was a fundamental connection between disease and sin. Disease was punishment for sin.' This interpretation is corroborated by Nelson and Rogers (1994) adding that during the Middle Ages, the care of the sick and those in need was a familial affair.

Without going too far into medical history, it must be noted that the scientific basis for water and sanitation issues in the beginning of the nineteenth century, was based on the so-called *miasma* theory on infectious diseases. The theory held that odors and vapors emitted from contaminated sources were the roots of sickness. Although the theory eventually was proven wrong it had a strong impact. When the first large European cholera epidemic reached Sweden in 1834, the miasma theories dominated, and it was not until the 1880s that modern bacteriology had its breakthrough. Thus the arguments used in the debates on water and sanitation were most often based on the miasma theory. The aim of 'hygienism' was abolition of dirt and filth, a standpoint that found no real poof of the scientific findings of bacteriology, but still an effective strategy in combating infectious diseases. Thus the Sanitary movement, despite its erroneous cause and effect theory, by identifying causes for disease, such as water, sewage, and bad housing, paved the way for bacteriology. Proponents for sanitation as a remedy for illness were on the right path of the wrong medical reasons.

In a recent investigation of provincial doctors in Sweden during the nineteenth century, historian Drakman (2018) convincingly show that the miasma theory lingered on until the end of the century while the doctors' perceptions of dirtiness, impurity, and causes to illness changed radically between 1865 and 1900. During an earlier period 1820–1865, the doctors explained morbidity with the geographical, climatological, and meteorological conditions of the district; the health of the local population was determined by the place they lived:

> 'But in the years 1865–1900, people's morbidity was rather linked to the cabin they lived in than the landscape where they grew up. The focus shifted to the filth and crowding that the provincial doctor saw around the patient's bed. The provincial doctor's attention narrowed. At the beginning of the century, human-borne contagion was explained by the arrival of strangers, but by its end it was no longer travel that was a medical problem, but people living too close together.'

However, it is difficult to assess this change in focus toward dirtiness in the private homes since the living conditions of the local population did not change during the first half of the nineteenth century. Cleanliness habits for common people most likely were the same for the whole period. Rather it was the medical doctors' perception that changed, and they reinterpreted dirtiness as the root cause for illness. Drakman claims that from the middle of the century, everyday dirt came to be regarded as miasmic and human refuse and overcrowding gradually became causes of disease. This was not at all the case in the early 1800s. Common dirt had not been identified as smelly miasma but related to lack of orderliness and neatness ('Politi'). Illness was now reinterpreted as something that came from within bodies and disease transmission took place inside the home. This change of attitude is also evident in the Health Care Act of 1874 which obliged municipal committees to deliver an annual report to the provincial doctor with information about the common people's dirtiness, overcrowding and if they lived too close to livestock. Also how people handled feces, saliva and other bodily secretions became central in disease avoidance (Nelson & Rogers, 1994).

6.4 THE SANITARY MOVEMENT AND THE HEALTH ACT OF 1874

It is obvious that the contextual factors related above, also in a European perspective, influenced the so-called Sanitary movement which had a huge direct impact on the management of municipal water and sanitation. The Sanitary movement can be seen as the final articulation of publicness in response to the contextual factors.

In the first half of the nineteenth century, earlier ideas on the mercantilist virtues of population development were challenged by the high mortality rates, the poor health of the population, especially in towns, and the many cholera epidemics. The industrialization process required a healthy labor force. Loss of labor productivity due to bad health became an economic problem. Thus, authorities, in Sweden and all over Europe, started to take a renewed interest in population development and sanitary issues became an important tool. The Sanitary movement originated in England in the beginning of the nineteenth century and the new perceptions of health and sickness related above were combined with the realization that the industrial workforce was a valuable production factor, which made it important to keep the workers healthy. Added to these insights was the realization that health and modern technology had to be joined together and true progress in public health was only to be achieved if water and sanitation was managed as a proper infrastructural system. I will not go deeper into this aspect, but it must be noted that Chadwick and the early Sanitary movement had an outspoken holistic view on the reuse of human wastewater and latrine. The piped sewage system with its connected water closets were meant to transport fecal matter and waste to farmlands outside the city where farmers would pay for them and use as fertilizers. This business model would finance the sanitary improvements in the cities and at the same time benefit agriculture. In the words of Chadwick, using an ancient metaphor, the recirculation of nutrients would complete the circle and realize the Egyptian notion of eternity by bringing 'the serpent's tail into the serpent's mouth' (Hallström, 2003). This holistic view was also central in Circulus, a French ideological movement advocating recirculation under the leadership of the utopian socialist Pierre Leroux (1797–1871). The idea was that human excrement should be collected by the state in the form of a tax and used as fertilizer and thereby end world hunger (Simmons, 2006).

Edwin Chadwick led the work preparing the influential *Health of Towns Enquiry* of 1844 in Britain. The report stated a clear connection between poor health and poverty and argued that sickness and epidemics like cholera spread via bad smells from putrefying matter (*miasma*). Public health should be prophylactic, and the solution was technological systems (piped water and sewerage) using water closets. British public health and modern WS technologies were influential across the Western world for decades: 'What was essential here was the successful British linking of public health *and* technology' (Hallström, 2003). Chadwick and the Sanitary movement changed society's perceptions on disease and poverty from something that was blamed on individual character. Instead they claimed that poverty was often the consequence of disease which had its roots in sanitary conditions for which the individual could not be held responsible. He strongly argued for preventive actions and that it would in fact be economically sound to prevent disease. Rosen (2015) claims that Chadwick 'proved beyond any doubt' that disease was caused by filthy environmental conditions, and that the remedy was better water provision and sanitation/ drainage. Thus, the problem of public health was reoriented and declared to be an engineering rather than a medical problem. Filth was raised to the status of an important public enemy (Rosen, 2015). According to Binnie (1981), it is evident that Chadwick had a clear understanding of the systemic and socio-technical nature of WS projects and that he argued that a water and sanitation system must be based on the science of engineering '... of which the medical men know nothing ...' The extension of water supply to every house and room, and even to the poor, was considered crucial and fear of a working-class revolution was evident. Chadwick's Health of Towns Enquiry and other studies claimed that the lower classes did not have any real insight or capacity for an orderly way of living. They had to be led on a new path which promoted personal cleanliness, clean housing, and clean cities including introduction of water supply, which according to Chadwick '... had proved to induce much sounder ways of living among the working class' (Hallström, 2003). Following this line of thought, indoor fittings were the best solution to sanitary issues; British engineers like Hawksley and his Swedish colleague Richert strongly argued against public wells and public taps since they slowed down the installation of internal water taps, which was seen as the best solution when it came to sanitary and moral improvement of the poor. Arguments were also made in favor of internal fittings because public pumps and water taps were '... places for inappropriate social behavior, for example, gossip, especially among women' (Hallström, 2003). These types of gendered assumptions are quite common in the history of technology. The introduction of the telephone, for example, includes lots of prejudiced statements, not at least from the telephone companies, on 'improper' female use of the new infrastructure. It is obvious that hygiene and cleanliness had physical as well as moral implications. Since this book mainly discuss the articulation of publicness on a quite abstract level where male engineers and decision makers were at the forefront, gender issues are largely hidden. However, it is evident that the introduction of piped water and sewage had an impact on gender equality and women's life situation. Not at least as it most often was their job to fetch water and manage laundry and inner sanitation. Similar gender aspects can certainly be found in street keeping and, for example, in the introduction of street lighting (Andersson-Skog, 1998; Fischer, 1988).

Even though in Norrköping, Stockholm and in other towns, sanitary concerns were the most common, one must keep in mind that these were not the only motives behind piped water. Other main arguments were street cleaning, public baths, industrial needs and not the least, improvement of municipal fire protection. Fire protection was of course central to cities crammed with wooden houses that easily caught fire.

It was against the background of cholera epidemics, the valorization of workers health, the high mortality rates, the new perceptions of health and sickness and inspired by the British Sanitary movement that the creation of Sweden's first comprehensive public health legislation began. The Health Act can be seen as the specifically Swedish articulation of publicness responding to the challenges from the contextual factors discussed above and as a direct operationalization of ideas from the sanitary movement on a national scale.

Advances in technology were of course also very important but technology is not a silver bullet; water and sewage systems were not new to the world. What was new, in Europe as well as in Sweden, was the creation of new municipal organizations with the power and resources to afford and manage these expensive undertakings. This can be seen as a successful articulation of publicness which gave political leverage and organizational capacity. In Sweden, as discussed, this organization, the Municipal council, was established through the Municipal Reform Act of 1862, which also gave the towns and municipalities the right to tax all citizens and take loans. The reform put health care in the hands of the municipality and made every town responsible for water and sanitation. But already in the 1830s, when the first cholera pandemic occurred, legislation was put in place both for quarantine regulations and, as mentioned, for the obligation for local parishes to report cases of cholera to the medical authorities. In towns, special committees were established which became the foundation for the establishment of compulsory local health boards after the approval of the Public Health Act of 1874. According to Nilsson and Forsell (2013), these early health committees can be seen as a prelude to the whole movement which led to the municipal reform of 1862 and the following city charters. The health boards of 1874 were put up to oversee, among other things, water, sewage, cleaning, garbage disposal, burial sites, housing, disposals, food handling, animal husbandry and industries.

Nelson and Rogers (1994) characterize the new act like this:

'The Public Health Act of 1874 was a piece of comprehensive legislation dealing with the entire range of what we today would include in the realm of health and environmental policy, and including measures directed toward the community as well as toward the individual. It was innovative in the sense that no attempt had ever been made to pass legislation covering all aspects of preventive health care.'

As mentioned, the Health Act of 1874 was heavily influenced by Chadwick and the Sanitary movement and the committee that worked on drawing up the act frequently referred to studies from Chadwick and associates that tried to estimate the economic benefits of a better public health program by calculating the number of man-hours gained in production and the reduction in costs of medical and hospital care. The conclusion was that improvements in water supplies and sewerage systems would lead to monetary gains for society as a whole (Nelson & Rogers, 1994).

doi: 10.2166/9781789063981_0065

Chapter 7
Modern water and sanitation

Before I move on in the history of modern drinking water and sanitation, which eventually led to the establishment of a modern water and sanitation system, I will shortly account for the early nineteenth-century's history of latrine collection and street cleaning. Moreover, as a contrast to the history of modern water provision, I will also briefly discuss water legislation excepting drinking water from the first half of the same century. Once again, I want to point out that the following account of modern infrasystems has an unfortunate center of gravity in the history of Stockholm, mostly of practical reasons, because earlier research has mainly focused on the capital city. Nevertheless, it is also quite clear that Stockholm in many cases was a forerunner or at least part of the forefront in WS development. What happened in Stockholm and a few other larger towns such as Malmö and Gothenburg soon spread all over Sweden in a similar fashion.

7.1 LATRINE COLLECTION AND STREET CLEANING IN THE EARLY NINETEENTH CENTURY

As mentioned above, street cleaning and gutters, i.e. outer sanitation were seen as public concerns, and connected to street keeping, what I have called *general* street maintenance, since the Middle Ages, although it seems that the authorities were not very successful in enforcing the many regulations. In the first half of the eighteenth century, a praxis was established meaning that property owners kept the street clean outside the plot and the city authorities managed street cleaning on open spaces and public squares. Excrement and latrine management, inner sanitation, on the other hand, was mainly seen as a private matter. Even though public authorities could complain about stench and flows of excrement, if these substances were kept within the confinement of the private property, they were not able to intervene. However, at the end of the nineteenth century, these areas of sanitation had been firmly moved under the town authorities. It is evident that the articulation process of publicness in especially latrine collection and, somewhat later, in street cleaning predates piped water and piped sewerage in the latter half of the nineteenth century (Hallenberg, 2018).

From around 1800, the meaning of the term sanitation changed. As mentioned, earlier sanitary measures that had been associated with 'Politi,' that is good order, cleanliness, and tidiness, became gradually associated with human health and wellbeing, a matter that increasingly became a public concern. Later, I will continue the discussion on how piped water and sewage to a higher degree than before became public responsibilities influenced by the advent of these health and sanitary ideologies. However, health and sanitary discussions of course also influenced latrine collection and street cleaning in the beginning of the century.

Finnish historian Henry Nygård (2004) describes this *prophylactic* strategy on cleanliness as a '... social purification process where dirt, infection and sin were just different expressions of the same evil principle. Expressed in social terms, it meant mastering not only uncleanliness and overcrowding, but also poverty, destitution, alcoholism, and prostitution, and all those circumstances which broke down the health of the poorest and left them susceptible to suffering and misfortune.' Thus, the prophylactic view on health and sanitary issues first appeared in latrine management, somewhat later in street cleaning, and gradually grew stronger in influence in the 'Sanitary movement' which directly affected the development of piped water and sewage.

Regarding latrine management, the so-called pit-system lingered on for a long time even if it was formally abolished in the 1880s. The scale of operations in latrine collection is hard to grasp. In 1880, there were almost 35,000 such barrels for latrine in Stockholm and only approximately 6000 were located indoors. On average, 4.41 Stockholmers shared each barrel. It was the house owner's responsibility to get the barrels transported and emptied. But because of the relatively high cost for the 'bidding,' it often took a long time between pick-ups which resulted in sanitary problems with overflowing and leaking barrels. In 1880, more than 274,000 collections of barrels were executed in Stockholm, and they were emptied into barges at in the harbor or taken by rail delivered to farmers outside the city where the latrines were sold as fertilizers. 'The system was functional in terms of circulation but hardly in terms of hygienic conditions. From the latter aspect, the water closet (WC) system could of course offer better solutions. Against this background, it is understandable that Strindberg, in his first encounter with the water closet, thought he had experienced something magical' (Jakobsson, 1999).

In March 1849, a committee was set up by the financial board in Stockholm to investigate how to handle these arrangements. The committee suggested that public authorities should take a firmer grip and wanted to get rid of the traditional ways of managing latrine collection. After some turns in the debate, where a publicly hired contractor was appointed, and very soon failed, in 1859 the municipality finale took charge by creating a municipal sanitation organization.

According to Hallenberg (2018), latrine management in Stockholm was organized like this before the 1860s:

- Up to the first half of the eighteenth century: as communal inconvenience executed by individual property owners.
- 1738: municipally organized contracting.
- 1760: communal inconvenience and property owners pay contractors.
- 1774: contractors hired by the municipality.
- 1800–1850: communal inconvenience (again) and property owners pay contractors.
- 1849: full contract for 10 years, paid by the municipality (the contractor fails).
- 1859: municipal organization: Stockholm Sanitation Works.

As a side note it can be mentioned that like common wells, bigger towns also had common privies to some extent. In Stockholm, according to a regulation in 1792, it was stipulated that public 'convenience facilities' were to be put up and to be managed by the contractor responsible for latrine collection. In the 1840s, there were around 15 public convenience facilities in Stockholm and in the 1870s, the city increased the number by allowing private individuals to rent privies and make them available to the public for a certain fee. Around 1885, there were 17 public commercial privies called 'cabinets' and at the same time around 70 public urinals in the city (Blomkvist, 2023a).

When cholera broke out in Stockholm in 1853, and claimed around 3000 lives, an invigorated debate began about the city's sanitary conditions. In 1859, the financial board decided that the city should also take over street cleaning alongside latrine management (Dufwa & Pehrson, 1989). However, this initiative was not very successful, and in 1875, a street cleaning company (Gaturenhållningsbolaget) owned by the city was established, billing the property owners for its service. Gradually, between 1895 and 1902, the city took over street cleaning in its own municipal organization. In 1902, street cleaning and latrine collection were merged in the Stockholm Sanitation Works (see above) when the street department of the municipality got the responsibility for both (Drangert & Hallström, 2003).

The chronology of street cleaning:

- Up to the first half of the eighteenth century: as communal inconvenience executed by individual property owners. Many different forms of contractors were also tried.
- 1859: the financial board takes charge.
- 1875: the street cleaning company.
- 1895: municipal organization established.
- 1902: excrement handling and street cleaning merged in the Stockholm Sanitation Works.

According to Hallenberg and Linnarsson (2016), the articulation of latrine collection as a public responsibility spearheaded the articulation of publicness in Stockholm and broke down the resistance toward a public, tax-funded organization. However, it still took quite a long time to introduce municipal street cleaning. Although the financial board advocated public street cleaning very early, it remained an individual responsibility of the property owners until 1902. It is in fact still a small part of the responsibilities of every house owner in Swedish towns to clean the pavement in front of one's own plot (Hallenberg, 2018). The difference in the articulation of publicness in street cleaning and latrine collection, where street cleaning became a public responsibility later than latrine management, is a bit paradoxical. Street cleaning and outer sanitation had since a very long time been an area where public authorities tried to articulate publicness. Latrine management and inner sanitation, on the other hand, were seen as an area outside the realm of the public. This paradox can probably be explained by two reasons: first, the prophylactic view on sanitary and health issues, for obvious reasons, was much more prominent in latrine management which had more acute negative externalities such as foul smell. Latrine collection also became an increasingly problematic area when the sanitary issues were put high on the agenda and the former private sphere of inner sanitation was opened to public intervention. Second, which was discussed earlier, street cleaning was connected to establish practices in road and street management where road maintenance was considered as tax payment in kind. If regulations for street cleaning were to be changed, it would disrupt an important principle behind tax withdrawal. This entanglement between the road and street sector and taxation gave the road and street system an institutional inertia and path dependence which resisted change and made the field very hard to modernize.

7.2 WATER LEGISLATION EXCEPTING DRINKING WATER

In the first Swedish national legislations on water, the non-public character of drinking water is reviled by the fact that drinking water was not mentioned at all. In the Act on Water Rights from 1865, the damming wheel for hydropower, the flotation wheel and the ditching wheel were the only water aspects addressed. The subsequent Water Rights Ordinance of 1879 made additions to urban water intake and fishing rights. However, nothing was mentioned about drinking water provision, groundwater extraction or domestic water from watercourses. Thus, contrary to the situation in ancient Rome, in Swedish legislation drinking water was generally treated as a private good and a productive resource rather than a public good (Drangert, 1991).

Swedish water legislation was, just as for civic roads, based on rules recorded in the landscape laws from the thirteenth century. They in turn built on ancient customary law based on the medieval village organization. In the 1734 law, water regulation was introduced into the Building Code. The rules thus applied partly to how one could appropriate water through various companies such as mills and partly to ditches, dams, and other projects that aimed to protect from the harmful effects of water.

Modern water legislation had its beginnings in the Ditching Act of 1879 and concerned so-called *defensive* undertakings, that is removing excess water from the fields. But its purpose was anything but defensive. The state realized that the cultivation of the kingdom required vigorous measures. In the Ditching Act, there were regulations for setting up ditching companies and rules for the formation of a Common Pool Resource (CPR) organization for this purpose. In this water related area, outside drinking water provision, inspiration clearly came from the methods used for management of common resources in the villages and both land drainage and civic roads were taken care of in the same manner. From the end of the nineteenth century until the 1960s, between 30,000 and 40,000 CPR organizations for land drainage were set up, of which around 10,000 can be said to be active today. In the Water Act of 1918, the legislation took on the *lucrative* water projects. An obvious reason was the state's desire to appropriate hydropower for the expansion of the electricity grid. Within both the defensive and the lucrative water rights, it is quite clear that the state, so to speak, perceived a need to step up the regulation because the water rights had become a national concern. The expansion of agriculture, industrialization, and the project to electrify Sweden made water a strategic resource in the building of a modern society (Blomkvist, 2010; Landin & Henrikson, 2022).

This is a chronological summary of the most important laws and regulations, based on Christensen (2003, 2015):

Water legislation excepting drinking water:

- Landscape laws (thirteenth century).
- Gustav Vasa's order (sixteenth century).
- House inspection arrangements (eighteenth century).
- 1734: Building Code: Swedish water legislation, just as for the individual roads, built on rules recorded in the landscape laws from the thirteenth century. They in turn built on ancient customary law based on the medieval village organization. In the 1734 law, water regulation was introduced into the Building Code (not mentioning drinking water).
- 1879: Ditching Act (based on 1915 Ditching Committee): defensive water projects
- 1880: Water Rights Ordinance: lucrative water projects.
- 1918: Water Act, 1918: 523 (included a few regulations on drinking water; replaced the 1880 Water Rights Ordinance: water pollution regulations were introduced in 1942).
- 1983: Water Act, 1983:291 (replaced the 1918 Water Act).
- 1998: Act with special provisions on water activities (replaced the 1983 Water Act).
- 1986: Nature Conservation Act (amended land drainage in 1992).
- 1999: Environmental Code.

7.3 THE ERA OF BUILDING PIPED WATER SYSTEMS

The first suggestion for a municipal system for piped water in Stockholm was put forward in 1851 (Cronström, 1986; Höjer, 1967). At the same time, the Swedish society of physicians started a campaign for modern water provision motivated by an investigation of mortality rates. The doctors were also inspired by a committee set up in Copenhagen with the same purpose and, perhaps most importantly, by the sanitary reforms described in Chadwick's *Health of Towns Enquiry* (Thelle, 2019).

In the campaign, hygienic, social, medical, and national economic advantages of piped water were pointed out, not to mention its importance for firefighting. Those who suffered the most from the current conditions were the poor, who lived far away on the wells in the city center. Women and children in poor families had to spend time and a lot of work fetching water, which they then had to save and use for a long time. In this way, much suffering had been caused in the form of back pain and leg injuries on the one hand, and stomach and intestinal diseases on the other. The rich could always hire people to carry water. The views of the Society of Physicians were cited by Wilhelm Leijonanccker who was commissioned to design the water supply system.

Despite of a quite strong internal resistance, this opinion led the parishes, the magistrates, and the Eldest of the Burghers to pay half the cost for a study trip to England and Germany for Leijonanccker. The other half was covered by the fire insurance company of Stockholm, which shows the importance of firefighting as a motive for piped water systems. In June 1853, Leijonanccker delivered his plan including maps, design, and drawings for a piped water system.

For the future organization, it was important that Leijonanccker decidedly advised against private companies building and operating water mains, sometimes in competition with each other, which was often the situation in England. His strong opinion was that profit interests should be completely kept away from such activities. In April 1854, the committee appointed to evaluate the proposition agreed with Leijonanccker that a water main should be installed, owned, and operated by the municipality. In the pro-arguments, social conditions, and sanitary issues dominated the rhetoric. It must be noted that later, piped sewerage also came under municipal ownership in most towns and municipalities. In fact, sewerage was considered a clear *natural monopoly*, even more so than piped drinking water, which made municipal ownership the obvious choice.

However, the project was far from secured. Some of the more peripheral parishes objected strongly. They did not want to pay for something that only rich people in the city center would be able to use and a service that they would have to wait for years to get. The resistance came mostly from the lower end of the Burghers, small artisans, and traders, and from not so well of property owners. Nevertheless, after quite a few complicated rounds of discussions and voting in the committee, the pro-water system side won with 42 votes against 16 and the qualified majority needed was reached in 1854. In short fire protection, social and sanitary arguments won over economic and financial misgivings. Further negotiations were needed to get the parishes on board and an investigation of the water in wells was added to the argument. It was claimed that a cholera outbreak in one parish, Kungsholmen, which was particularly opposed to the water system, was related to foul well water (a proposition that could not be proved though). A special board to manage the water works was appointed in 1856 and Leijonanccker's proposal was approved in September 1857, under the condition that the famous British expert Thomas Hawksley was in favor. Finally in February 1858, after Hawksley's approval, the work could start, and from November 1861 the public could collect tap water from six different collection points and the expansion of the system started. Connection to the water supply network was initially slow and during the late 1860s the common wells were still widely used. But these were said to be contaminated and gradually more people connected to the network (Anderberg, 1986).

Other towns that built water pipe systems during the same period included Karlskrona 1862–1864, Jönköping 1865, Malmö 1866, Gothenburg and Skövde 1871. During this decade, water pipes were also introduced in Kristianstad, Lund, Landskrona, Linköping, Uppsala, Askersund, Gävle, Uddevalla, and Sundsvall (Tjulin, 2002). Population size was of course an important factor, but also smaller towns built modern water systems. Jönköping started to build its system as early as 1864 and was the third city in Sweden. Norrköping, on the other hand, which was the third biggest town after Stockholm and Gothenburg got its piped water supply 10 years later. Other larger latecomers were the towns Gävle, Helsingborg, and Örebro. Furthermore, the motives behind piped water varied. In Karlskrona, the water pipeline was built by the crown to primarily meet the needs of the naval base. In 1859, the crown approved a proposal presented by Leijonancker and the work was finished in 1863. In 1897, the crown handed over the water pipeline to the city of Karlskrona. In Jönköping, the initiative came from the State-owned railroad company in 1863 and as the city did not want to miss out on the railway connection, construction started and the facility was completed in 1865. The motives in Norrköping were connected both to sanitation and the fear of fire. The first proposal for a water supply came already in 1826 but it was not until 1872 that the actual project was launched, this time under the leadership of Josef Gabriel Richert. In Gothenburg, the older proto system was replaced between 1868 and 1871 primarily because of water shortage and pollution of local water sources such as the river (Göta älv.). Several proposals were launched for piped water and sewerage during the 1850s and 1860s, but they were never realized. Finally, in 1864, the City Council hired Leijonancker to build a piped water network. However, the job was taken over by J. G. Richert, and around 1871 the system was completed. In Linköping in 1870, a proposal was submitted to the city council for building a water main. The motives were sanitary and economic advantages and improved fire protection. As one of the few examples, the city gave a private company the task of building the piped water system and Linköpings Vattenlednings AB was formed with the governor and mayor among the board members (Andersson, 1971; Bjur, 1988; Hallström, 2003).

It must be noted that the initiative to build a piped water network in Stockholm was launched 10 years before the acceptance of the Municipal Act of 1862 and that the system in Stockholm was ready for operation 10 years before the Health Act of 1874. The same early start can also be seen in a few other places which is a strong argument for the prevalence of the specific contextual factors discussed above. Publicness in drinking water was articulated stronger and stronger in the first half of the nineteenth century and the Municipal Act of 1862 and the Health Act of 1874 codified a changed perception that was already a fact. As mentioned, many other towns followed suit and Sweden already had functioning water works in approximately 10 towns by 1874. In 1875, 14 of Sweden's cities had water mains and in 1909 the number was 65 (Söderholm, 2007).

According to Tarr (1996), nineteenth-century cities moved from localized and labor-intensive service arrangements for water supply to more capital-intensive systems using distant sources, because of four reasons in addition to population increase (all of them applicable also in Sweden):

- Water from local sources used for household purposes was often contaminated, tasted and smelled bad, and was suspected as a cause of disease.
- More copious water supplies were required for firefighting.
- Water was needed for street flushing at times of concern over epidemics; developing industries required a relatively pure and constant water supply.
- In addition, rising affluence in the nineteenth century undoubtedly increased household demands for water.

As will be discussed in detail later, most cities constructed water mains before piped sewage. In these cases, sewage came as a response to the increased amount of water needing drainage. After the 1870s though, sewerage and water mains were built simultaneously to minimize excavation costs and the preferred method was the combined system (waste and storm water in one pipe).

To sum up, I again want to stress that the early development in piped water strongly indicates that the contextual factors discussed had a profound influence on the articulation of publicness and the development of water and sanitation and that the formal legislations were codifications of a more positive attitude in society toward publicness and interventions from the state and from municipalities in domains previously seen as private.

7.4 THE MOTIVES IN LEIJONANCKER'S PLAN OF 1853

Before moving on, I would like to take a step back and highlight a couple of points from Leijonancker's plan for the piped water system in Stockholm, which led to development accounted for above. I have already mentioned that Leijonancker used examples from glorious Roman times to advertise his message, but no ancient technical inspiration can be found in the report. It is clearly stated that modern WS technology originated in England (Blomkvist *et al.*, 2023).

First, it must be said that Leijonancker's report is an impressive piece of work. It is a complete design of a water system with very detailed technical and economic calculations, which I have no possibility to go into. As mentioned, the report was approved by Thomas Hawksley, one of the most prominent engineers of his time.

Second, I believe that if one wants to find a 'Swedish model' of sorts for WS infrastructure (although imported from England), Leijonancker's design plan is a good starting point. It certainly influenced many of the following towns building water, and later sewage systems in the decades after Stockholm.

Leijonancker's suggestion was motivated by these main points, covering all aspects that earlier research has listed as important components in water provision history:

- Concerns about the poor who lived far from water sources and on upper stories in the houses.
- More convenience for the better off citizens.
- Alleviate sanitary conditions in the whole town including street cleaning.
- Increasing the health standard of the population.
- A resource for water consuming industries.
- Firefighting gets more effective.
- Creates opportunities for public baths and washing facilities.

As a short note on my remark on a 'Swedish model' for WS infrastructure, I want to emphasize that even though Leijonancker was the first to design a modern piped water system and that his ideas certainly were influential, he was not alone. As mentioned, Josef Gabriel Richert was commissioned to design the water system in Gothenburg and later his son Johan Gustaf Richert became even more influential in water and sanitation. The correct way to attribute credit would probably be to say that the 'Leijonancker-Richert model' became the 'Swedish model' and the norm in system development.

Third, and lastly, I want to stress the profound influence of Chadwick's *Health of Towns Enquiry*. Leijonancker openly referred to the Sanitary movement and it reports on the health situation in England, comparing and adjusting the results to Swedish a context. Most of his general motivations for the piped water systems are blueprinted on Chadwick and the Sanitary movement.

In the report, he interestingly enough borrows the words of Thomas Hawksley in the *Health of Towns Enquiry* of 1844, the same words that later were used in the debate on piped water in Norrköping (Hallström, 2003):

'My own observations and inquiry convince me that the character and habits of a working family are more depressed and deteriorated by the defects of their habitations than by the greater pecuniary privations to which they are subject. The most cleanly and orderly female will invariably despond and relax her exertions under the influence of filth, damp, and stench, and at length ceasing to make further effort, probably sink into a dirty, noisy, discontented, and perhaps gin drinking drab – the wife of a man who has no comfort in his house, the parent of children whose home is the street or gaol. The moral and physical improvements certain to result from the introduction of water and water-closets into the houses of the working classes are far beyond the pecuniary advantages.'

As mentioned, following this line of thought, indoor fittings were the best solution to sanitary issues; British water and sanitation engineers like Hawksley (1858) and his Swedish colleague Richert (1869) strongly argued that public wells and taps would slow down installations of internal water and that they should be avoided. Furthermore, arguments against public water taps also included moral judgements based on a fear that these installations would turn into places were women would indulge all sorts of inappropriate behavior (Hallström, 2003; Hamlin, 1998).

Later on in the report Leijonancker adds what might perhaps be his own words:

> 'One might object that even greater mortality among the poorer class is not exactly a severe accident, a claim that can be defended if only people unable to work died, but the tables show that men between the ages of 25 and 40 often die, and from which, in poverty left families the proletarian class is mostly recruited. Now that sanitary measures undoubtedly contribute to increasing the middle age, the poor service can avoid many, perhaps most, such families'.

The vocabulary used by Leijonancker and other actors in the Sanitary movement is very unfamiliar to a present-day reader. The descriptions of the lower classes and especially its women are crude and seem quite cynical. One can wonder if the WS proponents really had any genuine concern for the citizens they spoke in favor for. The conclusion by, for example, Christopher Hamlin (1998) is that the motives among these sanitary engineers were not altruistic at all, and they did not campaign to improve health and mortality rates for the whole society. Their goal was to rescue working-class *men* from the perils of bad sanitation and raise the profits of industry through higher productivity (Hallström, 2003; Hamlin, 1998).

This conclusion evidently has some merits. However, it is difficult to interpret the words of historical actors and fully understand them today. The norms of how you talk about people or social groups have changed and it is easy to judge the actors according to your own moral yardstick. My own view on Leijonancker's words is that the question if he was genuinely and altruistically concerned for the poor is impossible to answer and even uninteresting. It is equally impossible to say that he was all together a cynic, serving the rising capitalists to get a healthy workforce. What can be said with some certainty is that he, and other actors in water and sanitation, adapted their language to an audience that listened more to economistic arguments than to sentiments of pity or social concern for the poor. If you wanted to change the articulation of publicness in water and sanitation and convince the authorities on the need for public involvement and public spending in competition with other pressing societal needs, you had to attune to the discourse of the day.

To conclude, sanitary and health issues were important, but not the only motives in Stockholm. Hallström (2003) also notes the many different motives behind public involvement in Norrköping. The sanitary advantages were given the most space, but the question of fire prevention was also considered one of the most important. Piped water supply would make fire protection '... so complete that a greater fire is nearly impossible.' Hamburg was mentioned as a model where the last big fire was pointed out as the motive for a large-scale water system. Fire protection was in Norrköping, as in Stockholm, also one of the main reasons accompanied by industrial needs such as water turbines. In Gothenburg, these motives were echoed and perhaps with an even stronger emphasis on financial arguments to enroll property owners, industry, and tradesmen in the project of public water provision (Bjur, 1988). Juuti *et al.*, (2009) claims that the primary use and the motive behind water systems in Swedish towns until the mid- and late 19th century was fire protection. During the era of piped water, devastating fires occurred in, for example, Ronneby, 1864; Karlstad, 1865; Gävle, 1869; Söderhamn, 1876; Hudiksvall, 1878; Karlskrona, 1887; Luleå, 1887; Umeå, 1888; Sundsvall, 1888.

7.5 FINANCING, MANAGING AND OWNERSHIP

In this section, I touch upon the central question in Linnarsson and Hallenberg's research on publicness: private or public ownership of municipal infrastructure. But I start with a short note on financing and management.

During the planning process in Stockholm, the managing board discussed how to design water tariffs. Individual meters were considered too expensive and technically insecure. The discussions ended up in a tariff based on the number of rooms in the property connected to the grid. Factories and commercial buildings were levied using meters though. Water for street cleaning, an important issue in the sanitary economy, was paid as a tax by the parishes in analogy with the tariffs for gas lighting. Furthermore, several standpipes were decided on where people not connected to the grid were able to fetch water for free. However, these public standpipes were not much appreciated by the managing board. They argued that too many standpipes would delay the preferred in-house connections and slow down the number of subscribers (see above). Fire protection was of course also an important issue, and it seems that this part of the water system worked satisfactorily (Höjer, 1967). Initial technical problems with water meters delayed their introduction, but in 1925 Stockholm introduced mandatory metering for every service connection, private and commercial, because a large share of produced water was unaccounted for. The new water meters led to a significant rise in income for the water works (Juuti *et al.*, 2009; Juuti & Katko, 2005).

Another example on tariffs is from Norrköping which in 1874 decided, inspired by Gothenburg, that the Waterworks Board should pay water and sewer service pipes up to the boundary of the private property, while the landowner paid for pipes laid inside the lot and inside the house (which is the same principle most municipalities use today). However, water for household use, except for horses and cattle, was free of charge, as was water for firefighting, municipal baths, and the cleaning of streets. 'Free' meant of course that the actual cost was paid by general taxation. Public buildings, hospitals, hotels, and factories had to pay a fee of SEK 3 annually for a tap in a workshop and SEK 6 for a water closet. In the small municipality of Eksjö, water for a room with a fireplace cost SEK 2 annually and if you had a water meter, 1000 l cost 20 öre (Andersson, 1971; Hallström, 2003). In Sweden as a whole, the question on how to finance the operation of the water works after they had been built was a troublesome issue in many towns. One strong argument was that if water had been extracted for public health reasons, and the most in need could not use it because of high fees, then nothing would have been gained. In Sundsvall and Gävle, for example, the municipalities decided to offer free water for personal use but took out a fee for commercial users. Eventually, partly because meters to measure individual water

Table 7.1 Construction of the first modern Swedish waterworks, 1860–1890.

City/years	Source	Fee/year
Stockholm 1858–1861	S	2 kr/room
Karlskrona 1861–1864	S	–
Jönköping 1864–1865	S	2 kr/room
Malmö 1861–1866	S	Taxation
Gothenburg 1867–1871	S	–
Landskrona 1869–1874	G	–
Lund 1872–1874	S	–
Norrköping 1872–1875	S	–
Linköping 1874–1876	S	3 kr/room
Uppsala 1874–1875	G	–
Gävle 1874–1876	S	–
Skövde 1875	G	1 kr/room
Sundsvall 1878–1879	S	–
Borås 1881	S	–
Vänersborg 1882	S	–
Härnösand 1882–1883	S	–
Helsingborg 1883	G	1 kr/room
Halmstad 1885–1886	S	–
Örebro 1885–1886	G	2 kr/room
Västerås 1887–1888	G	2 kr/room
Växjö 1887	S	–
Eskilstuna 1887	S	3 kr/room
Karlstad 1888–1889	S	2 kr/room
Mariestad 1889–1890	S	?

Notes: Adapted from Hallström (2003).
Surface water (S) and/or groundwater (G). The water fee is for household consumption at the first years of service.

consumption became safer and more affordable, all waterworks implemented individual tariffs. Gävle had free household water from 1879 to 1949 and was the last bigger town in Sweden to introduce water tariffs (Winnfors, 2008, 2017). In the conclusions, I return to the financial issues which are an important difference between roads/streets and water and sanitation. Roads and streets have almost always been financed by general taxation (in one form or the other) while water and sanitation been financed by a combination of taxation and user fees.

In Table 7.1, the development of the first 30 years of piped water is summarized (adapted from Hallström, 2003). The dominant source was surface water between 1860 and 1890. Only 7 of the 24 water works used ground water. Household fees based on number of rooms were used by 9 cities, one used a designated taxation (Malmö) and 13 did not charge anything, which again means that the bill was paid for by the collective of taxpayers.

Sourcing of raw water was of course important in the creation of the Swedish piped water systems. It seems to me that the water works were farsighted in their strategy to secure raw water resources. Already in the end of the 1800s, Malmö acquired land around Vormbsjön, Sundsvall did the same at Sidsjön and Stockholm bought 80–90% of the land around Bornsjön. The general idea was to secure the so-called 'rainfall area' surrounding the lakes to avoid pollution of the fresh water source. The same strategy has been used, and still is used, for groundwater sources. Adding to this, the water companies also acquired land for pipes and other installations. The land acquisition strategy is an important but under-researched area.

As mentioned above, Leijonancker was opposed to private ownership of the Stockholm water works. He motivated his standpoint with earlier experiences in Great Britain where commercial interests in his opinion had failed miserably. Thus, water works were considered public in stronger sense, including a demand on public ownership, in contrast to the earlier gasworks, that often were run by private companies in the beginning. In this respect, the piped water system, although local and under the municipality, was more like the state-owned national railroad network planned and built in the same period. Public ownership has lasted until this day and age. In the beginning of the twenty-first century, in tune with the prevailing zeitgeist, private ownership was investigated in the plans to replace the law on public water services (SFS 2006:412). But the investigation argued that water and sanitation services were a basic need and a prerequisite for a satisfactory standard of living. Water and sanitation was acknowledged as a *natural monopoly* and therefore not an area for private ownership and commercial interests. The municipal responsibility to provide was placed in Section 6, and the law was extended by adding environmental issues as a criterion besides health and sanitary issues (Söderholm *et al.*, 2022).

Juuti and Hukka (2005) and Juuti et al. (2007) present detailed facts about ownership in an international perspective from the establishment phase up to the year 2000. Their thorough historical comparison is too extensive to be fully accounted for. However, it can be noted that the first modern water systems often had builder–owner or concession models in European countries, and particularly in North America. In most cases, however, municipalities soon took over these water and sewerage systems. For example, in the early twentieth century, 93% of the systems in German urban centers were municipal, as were all the urban WS systems in Sweden and Finland. Thus in the 1920s, municipal ownership argued by Leijonancker in 1853 had become the dominating form of governance. This situation prevailed until mid-century when the Eastern bloc countries got state administrated systems after WWII. In the end of the 1980s, England and Wales performed a full privatization of the systems and in Estonia, a partial privatization was made in Tallinn in 2001. Since the early 2000s, water and sewerage services have been managed by a single utility in most European countries and cities (Braadbaart, 2009).

7.6 DID HEALTH IMPROVE WITH THE INTRODUCTION OF PIPED WATER?

As mentioned many times, the piped water system in Stockholm, and later in other Swedish towns, was to a great extent motivated by health issues in line with arguments from the British Sanitary movement. An interesting question for a historian would be if piped water, and later sewage, actually improved health. This is not as easy question to answer. It is true that cholera never hit Stockholm after the introduction of piped water and it is also true that fewer children died from diarrhea, but to establish a clear causal relationship between piped water and better health is difficult. In the same period, the sanitary, housing and food standards got better which also can explain improved health. Historian Fredrik Petersson has tried to clarify these issues by investigating how piped water spread over Södermalm in Stockholm during the period 1880–1920. He follows the gradual expansion of the network street by street and compares it with statistics on mortality in child diarrhea. A clear positive correlation is found. However, the improved child health is not only connected to the accessibility of larger quantities of water. Water quality, which improved greatly during the period, is of course also a decisive factor. The authorities in Stockholm realized this early on and built up an impressive expertise in water testing of both chemical and bacteriological water quality (Petersson, 2005).

The relationship between mortality and sanitary measures was investigated in 1908 by the City District Physician in Helsingborg, Med. Dr Carl Lindman. He stated that both morbidity and mortality had decreased substantially, which he attributed to the introduction of water mains, sewers, epidemic hospitals, the cleanliness statute, and food control. In Lindman's study, it was shown that mortality in the cities was about 40% higher compared to the countryside in the 1860s, and by the beginning of the twentieth century, the differences were evened out, even though mortality had also decreased in rural areas by nearly 20% (Figure 7.1; Bäckman, 1984). Lindman stated that 'had the population

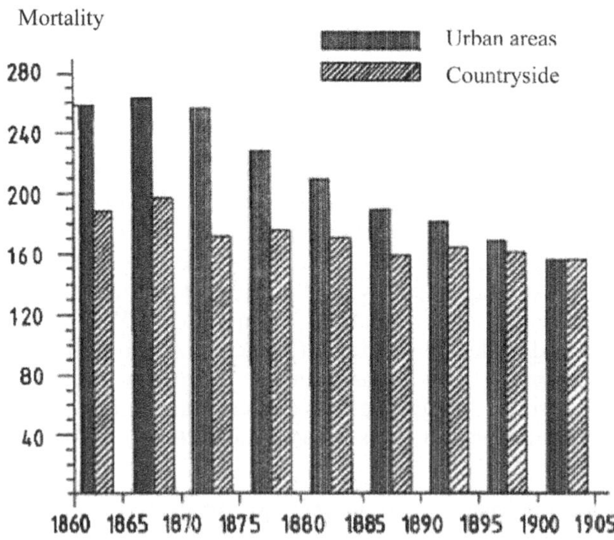

Figure 7.1 Mortality, calculated as number of deaths per 10,000 and year, in urban areas and the countryside. From Bäckman (1984), used with permission.

in Sweden's cities during the years 1901–1905 lived under the same hygienic conditions as during 1861–1875, then during these years 60,000 people would have had to pay for the decay.'

Given these numbers, it must still be noted that sanitary conditions, especially in the countryside, left a lot to which for, way up into the first half of the twentieth century. The already mentioned author and journalist, Ludwig 'Lubbe' Nordström, traveled around the country for the Swedish radio company in collaboration with the Swedish Medicine Board in 1938. His findings were not optimistic on the hygienic situation, and he coined the term 'Dirt-Sweden' (Lortsverige) to describe the filthy and unsanitary conditions people still lived under. Nordström's description of the miserable sanitary status became influential for the future expansion of water and sanitation systems. Interestingly, the large impact of Nordström's message was due to another infrasystem that by this time could reach most Swedes simultaneously, the wireless radio (Blomkvist & Kaiser, 1998).

7.7 INTRODUCING PIPED SEWAGE

In hindsight it seems a bit strange that water and sewer pipes were not built simultaneously. Presumably the towns did not dare to venture into yet another large-scale undertaking. The prevailing view was that not before the water main had reached its full extent, and only then, the surplus funds could be '... used for the construction of pipes for receiving and discharging overflow or impurity water' (Cronström, 1986). It was not until the middle of the 1870s that sewer networks got any significant expansion, first in Stockholm and then spreading throughout the country. Many of the smallest towns did not introduce piped water until the 1920s and even later embarked on a sewage network. However, there is one interesting example of the opposite. Luleå, a small city in the far north of Sweden, planned a network piped sewage before water mains. In Luleå, the large river running through the town was considered sufficient for drinking water and sewage issues were seen as more pressing (Söderholm, 2007). The example of Stockholm building underground piped sewage was soon followed by other towns in the following decades: 1860s, Gothenburg; 1870s, 8 towns; 1880s, 21 towns; 1890s, 23 towns; and 1900s, 18 towns.

Seen as an infrasystem, modern sewage, in contrast to earlier privy vault–cesspool arrangements, were capital rather than labor intensive and operated in an almost automatic fashion with less need for manual work which obviously improved sanitary conditions solving both collection and transportation when moving waste to a distant water curse. The main reasons for piped sewage according to Tarr (1996) were (referring to the development in the USA, but still applicable in a Swedish and European context):

- The capital and maintenance costs of sewerage systems would represent a saving over the annual cost of collection and cleaning with the privy vault–cesspool *system* (called *arrangement* in this book).
- Sewerage systems would create greatly improved sanitary conditions and result in lowered morbidity and mortality from infectious disease.

- Because of improved sanitary conditions, cities that constructed sewerage systems would attract population and industry and grow at a faster rate than those that did not.

Returning to Leijonancker and Stockholm, he was clear about the need for sewage and realized that his work was only half complete, and sooner or later sewer pipes must be laid down. In 1862, he put this argument to the board of directors of the water works and in 1864, he was hired to design a sewer system in parts of the town. In 1866, Leijonancker presented his plan, but it was rejected by the city council. The council admitted that piped sewerage would be beneficial for public health but argued that the project was to new, insecure, and unfamiliar in Sweden, and the decision had to be postponed until more knowledge and experience could be obtained regarding this type of systems. It was not until 1872 that the first sewage pipes were installed in the city center and the piped system slowly started to reach the whole town. These first pipes let out all the dirty water directly into the surrounding recipients. In the beginning, the sewage was mainly used for so-called gray water. The bucket arrangement for latrine collection prevailed for a long time and was not really gone until the 1940s and lasted longer still in smaller towns and in the countryside.

More generally, and in other Swedish towns as well, apartments in multifamily houses began to be equipped with kitchen drains in the 1860s, and the urine from the so-called floor exits (urine sorting toilets) was collected in a porcelain pot that was emptied into the kitchen drain (later a pipe for the urine was connected directly to the drainpipe in the floor). An economic reason for separating the urine was simply that the latrine barrel filled much more slowly which reduced the need for emptying (urine makes up about 90% of the volume of a person's excrement, however the value as a fertilizer also diminishes) (Drangert *et al.*, 2002).

Although the water closets were part of the debate and had many proponents in the 1890s, the WC was still an unproven challenge to more traditional service arrangements. The Stockholm city council and the finance board, who were now in charge of the water works and the piped system, were opposing WC while the municipal health board approved. This made the expansion of WC very slow. In 1895, Stockholm had only around 40 houses with this convenience installed, a number that had risen to around 1500 in 1904, the year when the restrictive attitude was somewhat loosened. However, it was not until the new sewage plan of 1909 that WC was allowed. When the ban was lifted, WC installations spread fast and the connection to the piped sewage system took off. By the 1910s, the water closet system had become an integrated and generally accepted system for sanitation, if not yet fully realized, in Norrköping and Stockholm, as well as in the rest of Sweden. The new National Health Care Act of 1919 confirmed the development and advocated that public water and sewer lines ought to be constructed all over Sweden (Drangert & Löwgren, 2005). Gothenburg also had its WC-debate in the 1890s. The town's leading actor in water and sanitation, J. G. Richert, expressed a clear opinion. He stated that the water flushing toilet was a 'hygienic axiom,' in other words something self-evident not needing further justification, and that counter arguments, like the fear of pollution, were exaggerated. In any case, the positive hygienic qualities of the WC outweighed possible negative effects, according to Richert (Bjur, 1988).

Bäckman (1984) using statistics from the Swedish Association of Municipal Engineers discusses how many of the cities' built-up plots were equipped with an underground sewage line, respectively equipped with a WC. Note that barrel collection could continue in many places in smaller communities and towns right up to the 1940–1950s. The towns and communities included in the statistics in 1922 (98) had a variation of built-up plots equipped with underground sewage pipes between 55% and 100%. Corresponding figures for plots with WC varied, with a few exceptions, between 0% and 30%. In 1927, most of the communities had less than 3 m of sewage pipes per person. Djursholm, the affluent suburb of Stockholm, on the other hand, had about 10 m per person.

The examples from sewage development in Stockholm below were mirrored in the general evolution in other larger Swedish towns and eventually also in smaller municipalities. The 1909 sewage plan for Stockholm presented by the head of the street department, C. J. Gimberg, included so-called 'cutting lines' that took care of storm water, and the building of new sewer outlets farther out from the city and its harbor. In 1925, it was decided to draw out part of the existing outlets to even deeper water. The plan of 1909 and the following adjustments did not really solve the problem with pollution of the water recipients. After 1925, the sewers were buried deeper and the discharges ended up below the water surface instead of directly on the surface, but this

Table 7.2 Construction of the first modern Swedish sewer systems, 1860–1890.

City/years	Sewer type
Stockholm 1866–1900	C
Gothenburg 1868–1888	C
Norrköping 1872–1874	C
Linköping 1874–1875	C
Uppsala 1874–1875	C
Sundsvall 1878–1879	C
Borås 1881	C
Härnösand 1883	C?
Örebro 1885–1888	?
Jönköping 1885–1886	?
Västerås 1887–1888	C
Karlstad 1888–1889	S
Mariestad 1889–1890	C
Lund 1890	C

Notes: Adapted from Hallström (2003).
Combined (C), separate (S).

did not help much. It became gradually more obvious that one could not any longer discharge the city's sewage completely untreated. In 1930, a new plan for Stockholm's sewerage was presented. It came to be known as the 'Sewage bible' with the official title 'Proposals for devices for the purification of wastewater in Stockholm' and it had a big impact also in other Swedish towns. It was the beginning of a large-scale planned sewage network in Stockholm including sewage treatment plants. However, the plan only aimed at purification of the wastewater from the inner-city area and a few suburban areas. For example, the newly built residential areas in Bromma got their and the city's first treatment plant in Åkeshov in 1934. A complete plan for the city's sewage treatment was therefore not presented in the proposal. The 1953 'General plan for the treatment of Stockholm's wastewater,' approved by the city council in 1954, became the starting point for the whole country's continued sewage planning. In addition, the plan presented the necessary collection lines, in the form of tunnels and pumping stations. In connection to the rearrangement of collection pipes for both the inner city and the outer areas, efforts were made to divert stormwater from streets and parks directly into the waterways to thereby reduce the inflow of stormwater in the sewage collection pipes (Cronström, 1986). As will be discussed later, which can be seen in Table 7.2, the domination of combined sewers was almost total during the first 30 years of sewer construction in Sweden (adapted from Hallström, 2003).

doi: 10.2166/9781789063981_0083

Chapter 8

Water and sanitation in the twentieth century

This chapter contains the most important traits in the recent history of water and sanitation (WS). It starts with the introduction of two additional contextual factors affecting the articulation of publicness from the first half of the twentieth century: first, environmental concerns due to pollution and second, a growing focus on sustainability.

8.1 TWO NEW CONTEXTUAL FACTORS

This section builds on a periodization of legislation in three *generations*, from the beginning of the 1800s until today introduced by Christensen (2003, 2015). Even though Christensen's focus is on *environmental* legislation, his timeline fits well in my account of modern water and sanitation history and the articulation of publicness. The different generations of environmental law are characterized by their most important aspects:

- *Sanitary aspect/health care aspect.* From the beginning of the 1800s, the culprit in the drama was not primarily emissions from treatment plants, but from individual drains that did not meet even basic purification requirements.
- *Pollution aspect/environmental aspect.* In the beginning of 1900, attention was paid to watercourse pollution and fish death. The supply of nutrients, mainly nitrogen and phosphorus, led to what we know today as eutrophication. Major culprits were leather processing industries and the cellulose industry.
- *Recycling aspect/sustainability aspect.* In the beginning of the 1980s, in the wake of the Brundtland report in 1987, people increasingly began to realize the extent that the availability of natural resources is not infinite. The need of recycling and reusing has been strengthened, which not least applies to society's exploitation of the element phosphorus.

Following Christensen (2015), laws and regulations in water and sanitation can be described, in more general terms, as originating in the healthcare statues of the 1860–1870s, followed by a focus on environment and pollution from the 1940s. The third generation was related to the national Environmental Code of 1999 which points to sustainable management of natural resources, reuse, and recycling of energy and nutrients (in the wastewater). It must be noted though that the term 'generation' can be a bit misleading as it signals three separated historical periods; the earlier generation dies when the new one is born. This is not the case, which is also acknowledged by Christensen, quite the contrary, the earlier factors are still very much alive. The next generation simply adds new contextual factors and today, these three generations of legislation have merged into the environmental code's regulatory system: 'Old regulations, which were based on the needs that existed in the old local community, must be combined with regulations aimed at reducing climate impact and that all water bodies in the EU must achieve good status. This means that what may look like a preconceived system of laws and regulations is in fact the result of many legislative actions over a long period of time.'

This first 'generation' would be equivalent to what I have described above: water and sanitation articulated as a public domain in relation to the five contextual factors culminating in the Sanitary movement and the Health Care Act of 1874. And, as mentioned, health motives for public involvement did not disappear. Some health-related regulations for wastewater handling were in fact already visible in the 1868 Ordinance Statute (1868:22), but it was in the 1874 Health Care Charter, which primarily applied in towns, that health and water and sanitation was firmly connected. Municipal authorities, through the mandatory Health Board, were obligated to arrange for the disposal of wastewater and a responsibility to supervise the quality of drinking water (also in the countryside). These regulations were transferred to the 1999 Environmental Code. The new Sanitary Charter of 1919 was also based on health and sanitary motives. It confirmed the articulation of publicness, by stating that modern sanitary measures included public water and sewage systems. The 1919 health care charter was replaced by the 1958 health care charter, which was in effect until 1983 when the Health Protection Act (SFS 1982:1080) came into force, which was in turn replaced by the Environmental Code of 1999. From the 1955 Public Water and Sewage Works Act, it was clear that piped water and sewerage in urban areas was a central part of municipal obligations due to health and sanitary reasons. The law clarified the municipality's obligations to provide for piped water and sanitation.

Simultaneously, as the health aspects of WS legislation were developing, the pollution of the environment because of untreated wastewater became evident. The emissions from individual drains as well as from industries, such as leather processing industries and the cellulose industry, clogged up water courses, killed fish and led to eutrophication due to an oversupply of nutrients, mainly nitrogen and phosphorus.

Environmental concerns were in fact discussed already in the so-called 1915 Ditching Committee preparing revisions in the 1879 Ditching Act, but it was until 1942 that environmental issues were included in water legislation, in an amendment in the 1918 Water Act. As discussed, legislators had so far focused on sanitary aspects, but the new rules aimed at protecting the environment.

This change in attitude came quite late concerning that political debates and scientists had since long called for measures on environmental protection due to water pollution from sewage. In 1956, requirements for further purification than only sludge separation was introduced and the 1964 rules were further tightened by introducing special protection for drinking water sources.

In 1969, the new Environmental Protection Act was introduced. This legislation was the first in water and sanitation that dealt with both health and environment issues in combination. However, there was still some regulation left in the 1958 health care charter, and later in the 1982 Health Protection Act, but through the 1999 Environmental Code, all regulations have been brought together.

The requirements in the Environmental Protection Act of 1969, together with the introduction of state subsidies, led to the initiative to build large treatment plants all over Sweden (discussed below). These initiatives meant that large-scale sewage related problems basically disappeared in Sweden, even though eutrophication is still a problem. The largest contributor concerning eutrophication is probably agriculture although individual, private sewage solutions also contribute with quite large discharges.

The second contextual factor affecting (late) twentieth-century articulation of publicness in water and sanitation (which would be Christensen's third 'generation') was of course *sustainability*. From the beginning of the 1980s, and especially after the publication of the so-called Brundtland report in 1987, the belief in infinite availability of natural resources was torpedoed and the need of recycling and reuse was emphasized. In water and sanitation, the rise of sustainability, and 'circular economy,' as driving forces for public engagement, has led to a focus on housekeeping off resources in sewage as well as in a focus on water conservation. It is perfectly clear that our present-day concerns for global warming effects and an upcoming climate crisis have strengthened the sustainability arguments and made it even more urgent to reuse and circulate resources as well as to reduce emissions and energy use.

In water and sanitation legislation, *sustainability* became the basis for the Environmental Code of 1999 which requires that consideration must also be given to the management of natural resources as well as the reuse and recycling of energy and nutrients. The intention was to create cycles of natural resources, with the aim of both reducing emissions and reducing the need to extract natural resources. The importance of these issues was strengthened with the Swedish adoption of EU directives. For example, the EU through the seventh environmental action program committed all its members to resource management and a circular economy.

Thus, in the post-war period and onwards, the public commitment was articulated through the environmental issues, which led to the massive expansion of large treatment plants. Then, in the late 1990s, publicness was articulated through resource cycles and sustainability. This discussion did not give so many concrete results but gained momentum during the first part of the twenty-first century with escalating problems because of global warming.

Because of the strong impact of environmental issues the WS system changed. The earlier focus on municipal engineering and system building in a practical and technical sense was influenced by other professional fields.

Civil engineering and health were superimposed by natural science dealing with the environment such as biology and chemistry. Water and sanitation were gradually divided into two parts with overlapping logics and different scientific foundations, where new actors seized parts of the problem formulation privilege.

8.2 WATER AND SANITATION ON A NATIONAL SCALE

When public water mains were built in Stockholm in 1861, the town got continuous access to water as a means of transport with the possibility for modern sewage systems, that is underground pipes in a sufficient slope to be self-purifying (which since the 1840s had been built in Europe's largest cities). The reason to put sewerage in pipes was that the old gutters were not sufficient to drain the cities anymore; waterlogging and clogged ditches causing stench and inconvenience became common problems. Based on simple hydrological principles, what came into the system must get out. Now, when even more wastewaters had to be disposed of, the old sewers were inadequate.

As Hallström (2003) notes:

'The sewers would transport dirt, "matter out of place," to its rightful place, the river, which, in its turn, would clean or dilute the dirt, or finally send it to the immense sea, where it would disappear. The rationale of the sewer system was thus perfectly logical to the actors, both in its theory and practice: it freed the city of harmful substances, and the running water of the river purified them, thus restoring the categories and the social order.'

As mentioned, in Stockholm, WC (water closet) connections were allowed to the piped sewage system in 1909 and the revised and National Health Care Act of 1919 confirmed the development and stated that modern sanitary measures included public water and sewer systems. From this year, it is evident that piped WS in urban areas was indeed a central part of municipal obligations (SFS 1919:566) and the pipe-bound infrasystem had become the norm in Sweden's national 'water and sewage strategy' (Drangert & Löwgren, 2005). During the interwar years, more and more towns built piped WS systems and after WWII, this development also reached the countryside (Söderholm, 2012). The motive was to include rural areas into the Swedish welfare society, and an important feature was a widespread expansion of WS systems starting in the early 1930s, with the help of national funding in the form of relief work issued by the National Unemployment Commission where water and sewage works often were subsidized with 90–100% of the total construction costs. However, this expansion happened under the general water law from 1918, which did not really include water and sewage systems. These areas were covered in the first national comprehensive legislation specifically addressing public water and sewage: the 1955 Act on public water and sewage facilities (SFS 1955:314).

In the state investigation preparing the Act of 1955, the investigators confirm my thesis that water and sewage has primarily been a private and not a public matter in a historical perspective:

'The water and sewage issues have only become the subject of more detailed regulation from the public's side only relatively recently. No uniform regulation at all has been achieved, but the regulations have been announced in various contexts in health care, planning and water legislation as well as certain other statutes ... In the countryside, the water and sanitation conditions are different than in the cities. It was not long ago that the WS question in rural areas was invariably considered the sole concern of the individual property owner' (Blomkvist, 2023a).

The investigation strengthens the impression of scattered regulations from many fields of law and a patchwork of many stakeholders. The act on public water and sewage facilities sets out to remedy this and to create a sound judicial base for water and sanitation systems. The conclusion was that older regulations such as the act of order from 1868 and the 1874 Health Act had failed to put water and sanitation in its right place. Not even the revised act of 1919 was free from criticism because it lacked clear regulations where a municipality can be forced to make WC facilities. The new law clearly stated that it was a municipal obligation to provide water and sanitation, and the state investigation argues that there is '... no reason to differentiate between sewage and water supply facilities with respect to the municipality's obligations. Within a city-planned area, the sanitary interests in general cannot be satisfied satisfactorily without common water facilities' (Christensen, 2015). This was the first time in the history of Swedish water and sanitation that publicness was clearly articulated in legislation together with a municipal responsibility to provide. The general principles laid down in the law on public water and sewage facilities from 1955 is still valid today although the law has been revised in 1970 (1970:244) and replaced first in 2006 with the new law on public water services and later in 2023 with a revised law on public water services (2022:1249).

Water and sewage systems grew fast in Sweden during the first half of the twentieth century; the total pipe length increased from about 3500 km to 10,000 km, and even more up until the 1980s. This physical growth was of course due to increased number of users, and also because of increased public investments and the forming of institutions and organizations supportive of the system, all contributing to system inertia (Cettner *et al.*, 2012).

The movement toward centralized municipal water and sanitation had its ideological roots in the growing conservation and environmental movement that viewed safe water as a human right and WS infrastructure as the solution. I agree with Söderholm (2013) who finds the connection between environmentalists and water and sanitation systems '... a bit paradoxical given that a key explanation behind this movement was the increasing wastewater pollution of Swedish watercourses from the increased use of WCs in urban areas in the early 1900s.' Untreated sewage caused serious problems and in the middle of the 1930s, only 10% of Swedish cities treated its wastewater. Thus, environmental concerns became important in Swedish society; although sewage was the main cause of the problems, it also became the solution.

The expansion of infrastructure in Sweden was also influenced by a municipal reform in 1952 that merged and reduced the number of municipalities from around 2500 to about 1000. This move gave the now larger municipalities increased planning capacity for undertakings such as centralized and big water and sanitation systems. An important tool for the state was grants and subsidies for the improvement of infrastructure. After the WWII, state grants were introduced on a larger scale for the expansion of the water and sewage pipes of cities and larger communities (SFS 1946:287; 1948:441). The maximum contribution was 75% of the construction cost. The state, via the Swedish Environmental Protection Agency (EPA), started a program for water provision in the countryside in the late 1960s aiming at environmental improvements. By now, the urban WS networks were largely completed. The state subsidy during 1970s primarily concerned improvements to protect surface and groundwater and the expansion of wastewater treatment plants targeted phosphorus reduction. Economies of scale lead many municipalities to connect sewage from areas in the city's peripheral surroundings to a central purification plant and since transmission lines would still be buried, it was advantageous to lay down water pipes at the same time, thus centralizing both water and sewage. The state subsidies ended in 1980, when the expansion was completed.

According to Cronström (1986), the general Swedish history of sewerage can be divided in these periods:

- 1910: no purification necessary.
- 1930: mechanical but not biological purification.
- 1953: mechanical and biological purification, but not nutrition salt separation.
- 1970: Mechanical and biological purification as well as phosphorus separation.

To sum up, there were ideological, environmentalist and economic motives for municipalities to push for centralized solutions. With the help of government subsidies and optimistic assessments of population growth, large, expensive sewage treatment plants had been built which meant that cost comparisons between a new, possible smaller, local plant and subsidized transmission lines to a central plant came out even. The investment made in a central solution had a decisive influence on the choice of subsequent investments. Also, in the 1960s and 1970s, actors shared mental images and visions of large scale, 'big is beautiful,' WS systems that had a very strong influence in the sector. Large-scale infrastructure was a new and exciting field for planners at ministries, the Swedish EPA, county boards and municipal politicians, and engineering consultants designing the large-scale solutions. The technical and the political elite had a common project. Furthermore, and as will be discussed later, these state initiatives and subsidies strengthened *vertical integration* in water and sanitation. The facilities were still local, and we did not have a national gridded system like in roads and streets, but from an institutional point of view, vertical integration still increased.

8.3 PATH DEPENDENCE IN WATER AND SANITATION

This section is a deepening (with some repetitions) of the discussion above, focusing on *technological* path dependence related to the building of combined sewer systems. As discussed earlier, when the piped water systems led more water into the cities, the problem of getting rid of the wastewater got more acute, especially after the introduction of the flush toilet (WC). Generally, in most cities in Europe, the USA and Sweden, piped sewage was introduced after the modern water provision. Urban sewers (gutters) were meant for stormwater, and they became *combined* sewers when households continued to use them after the installation of running water and when water closets were used. The problems with installing WCs were indeed discussed but the problems with its discharges still surprised most proponents of piped WS systems. Thus, due to the historical legacy of already existing technology, that is stormwater drainage, combined sewers for stormwater and wastewater became the norm.

Also in London, when the city constructed a sewerage system in 1858, based on the plan of Joseph Bazalgette, a combined sewage system was chosen. A separate system was rejected with the argument that storm water because of animal excrement was as much polluted as sewage and ought to be treated as such. Bazalgette, Chief Engineer of the Metropolitan Board of Works, designed a series of interconnecting sewers which carried the sewage eastwards away from the main centers of population to be dispatched on the outgoing tide. The Bazalgette plan also included the construction of embankments along large sections of the River Thames in central London. The embankments concealed the new sewers and acted as flood defenses (Halliday, 2001).

The combined sewers operated on the rationale of the theory of the self-purification of streams; that running water purified itself within a given distance. Up until the 1890s, this hypothesis seemed confirmed by existing methods of chemical analysis of water quality. Except for specific localities with severe nuisance problems from sewage disposal, municipalities resisted installing sewage treatment facilities that promised to provide direct benefits only to downstream cities and instead relied on dilution to dispel the worst concentrations of pollutants (Tarr, 1996). The belief in self-purification of running water was dominant also in Sweden. The influential physician Klas Sondén investigated water pollution in Stockholm in 1889 and concluded that sewage would not contaminate the recipients to any higher degree. He was more concerned that high tides in the Baltic Sea would lead to saltwater penetration into Lake Mälaren and thereby contaminating the freshwater reservoir. Others as, for example, J. G. Richert argued in the 1909 sewer plan that it would be desirable if all sewers were built so that they could be equipped with sewage treatment devices in the future. Sondén in 1910 turned against this view. He believed that it would be a long time before the sewage would cause any significant problems. Later, Sondén changed his opinion though and in 1930 advocated mechanical purification as it was '... inevitable to purify the sewage from floating impurities and sludge' (Blomkvist, 2023a).

In conclusion, the choice of combined sewers due to earlier technical design, that is street gutters and partly because of the belief in the theory of water self-purification, created a strong path dependence in water supply and sewerage. Nevertheless, the centralized system was also criticized. During the 1980s and 1990s, there was a heated debate on reuse of resources in wastewater, such as urine diversion and sludge as fertilizers in agriculture. It is worth noting that

Sweden was a pioneer in the field of resource-oriented solutions in sanitation. But these pioneering attempts failed, and the reasons were mainly connected to the inertia of the existing system. It was very hard to change the direction of a large infrastructural system to optimize the recovery of resources when it was originally designed to improve urban hygiene and to control water pollution (Söderholm *et al.*, 2022). The history of reuse and resource in sanitation is discussed by Vidal (2022) describing Sweden as an forerunner in resource-oriented sanitation, especially in urine separation techniques. There has also been a quite strong political consensus on the need for alternative sanitation solutions, although the practical implementation has been slow because of the reasons mentioned above (inertia and path dependence of the existing system).

8.4 PRE-MODERN AND OFF-GRID NEVER DISAPPEARED

In the evolution of an infrastructural system for water and sanitation in Sweden, there is one untold story. The account given so far might give the impression that the infrasystem is totally dominant, comprehensive, and covering all parts of the country. And it is certainly true that we can see a development from privately managed WS arrangements to public municipal infrastructure; the articulation of publicness is loud and clear. Nevertheless, it is very important to note that large parts of the countryside did not get access to piped water until way up into the 1970s, and water and sanitation systems were not really regulated until the first national water and sewage law in 1955.

As discussed in the chapter on roads and street history, there is a big portion of the Swedish road network, civic roads, that are managed by private property owners living close to the road according to the old mode of road keeping based on the pre-modern 'interest and utility' principle. But still, civic roads are quite well aligned with the public infrasystem and to some extent controlled by the state road administration. This is not entirely true in water and sanitation, where local level and off-grid service arrangements are less aligned to the public system. However, even if they are off-grid and not technically connected to the infrasystem, there are public rules and regulations mainly concerning environmental issues that must be considered.

Before going further on small-scale water and sanitation in Sweden, it can be noted that in Finland, cooperatives seem to be the preferred form of organizing decentralized water and sanitation. In an interesting article on alternative 'paradigms' in WS governance, Hukka and Katko (2009) point to Finnish experiences with its long tradition of cooperatives operating in rural and smaller urban settings for decades. There are approximately 1500 water and wastewater cooperatives in Finland. Cooperatives can be found also in Denmark, in the US and for a long time in Latin America. The actual number of cooperatives in Sweden is probably small. The preferred organization seems to be a joint property unit or community association (Samfällighet) just like in civic road keeping. A study focusing on the archipelago in Norrtälje, north of Stockholm (Roseen, 2020), claims that there are approximately 9000 of these community organizations in Sweden. Water communities, once formed, have the same powers as civic road communities. But the most important difference is that an individual property owner cannot force other residents to join the formation of a community organization for water or sanitation. In less densely populated places, everyone is free to cater for themselves if they can show that the individual arrangement for WS provision can be managed safely regarding health and environment. This legal disparity strengthens the thesis that water and sanitation has a weaker articulation of publicness than roads and streets. Water and sanitation is still to some extent a private matter and we can still see the legacy of the pre-modern unwillingness to intervene in the private sphere.

Returning to the question of pre-modern and off-grid, Sweden still has a large part of its population still depending on private arrangements. These, what might be called *pre-modern* WS arrangements, are still to a high degree present and consist of private property owners arranging for their own off-grid solutions. In Sweden's well-developed infrasystem with around 1.5 million properties connected to the municipal water and sewerage grid, almost 1 million properties (of which around 450–500 000 are leisure properties) are not connected (Blomkvist *et al.*, 2023).

The private character of water and sanitation is still a reality and public authorities accept their existence, especially concerning drinking water from private wells where regulations on water quality certainly exist but they are not as forceful as rules for private sewage arrangements. In fact, the public control of private wells is almost non-existent according to informants (Blomkvist, 2023a). The reason for the difference in attention and the view that drinking water quality is up to the individual, still not articulated as a fully public responsibility, is because private sanitation arrangements can possibly contaminate ground water sources, that is the well of your neighbor, and at the same time pollute nearby water courses and recipients: a higher risk for negative externalities. According to the environmental law (SFS 1998:808), the owner of a sewage facility has an obligation for proper maintenance to protect the health and the environment.

Although hard to estimate, around 25% of Sweden's one million small-scale private sewage facilities lack adequate treatment. This means that approximately 250 000 small-scale and private sewage arrangements possibly do not meet the requirements in legislation (Söderholm *et al.*, 2022). The Ordinance on environmentally hazardous activities (SFS 1998:899), Section 12 states: 'It is prohibited to discharge wastewater from a water toilet or densely populated areas, into a water area, if the wastewater has not undergone further purification than sludge separation. However, what is said in the first paragraph does not apply if it is obvious that such a release can be made without risk of inconvenience to human health or the environment.' Thus, discharging WC sewage without more extensive treatment than sludge separation is illegal only if the discharge takes place into a water area. Discharge of sludge-separated WC sewage is therefore legal if it is done to the ground, provided that no risk of nuisance arises. The crux of the matter is how to define the concept of *water area*. I have no opportunity or sufficient expertise to determine what is right or wrong in this debate. But it can be stated that opinions differ among experts on small drains. According to the authorities, 25% of the small sewers lack more advanced treatment. This issue is not under debate. But there is a lot of disagreement on the question whether this fact makes all of them illegal (Blomkvist, 2023a).

Instructions on how to build small drains were available from the 1960s onwards (Blomkvist, 2023a). They dealt exclusively with land-based facilities, that is purification using the soil. The focus of instructions from the 1960s to 1980s is on construction techniques and the chronology of public involvement looks like this:

- Sewage investigation 1950s: investigations concerning small sewage plants. The state's public investigations 1955:18.
- Royal Road and Water Works Board – small sewage works: Royal Road and Water Works Board 1962, No. 8, small sewage works.
- Small Sewer Facilities 1974 Guidance and description of technology for small sewers (infiltration and soil bed) from 1974.
- Special print from the National EPA's (SNV) publication 1974:15, 3rd edition. Small sewage plants – treatment of wastewater from individual properties.
- Nordic Council of Ministers and SNV 1985 infiltration of wastewater. Infiltration of wastewater – conditions, function, and environmental consequences. Nordic joint report: The EPA informs. Nordic co-production – EPA Nordic Council of Ministers, 1985.
- The Swedish EPA's general advice (1987) 87:6: small sewage plants – domestic wastewater from no more than five households (AR 87:6), often called 'the blue book,' apply to the design of conventional infiltration facilities and soil beds (has been replaced by Factsheet 8147, with the same content).

The requirements for off-grid sewage were rooted in health protection and the starting point was that sewage treatment was to be solved with the help of soil retention. A protective distance was specified, especially to wells. When the soil was unsuitable (too permeable or too dense), it could be a question of mini sewage treatment plants and sometimes a closed tank for WC (most used for holiday homes). During the 1980s, state authorities put the purification processes in the small-scale sewage under scrutiny and especially with the environmental law from 1999; the eutrophication problem was addressed in relation to off-grid sewage. This turn toward environmental issues was due to The Baltic Marine Environment Protection Commission, also known as the Helsinki Commission and its Baltic Sea Action Plan from 1974 where Sweden agreed to map and remedy nutrient leakage into the Baltic Sea to protect the marine environment from all sources of pollution. It was signed by all Baltic Sea coastal countries seeking to address the increasing environmental challenges from industrialization and other human activities that had a severe impact on the marine environment. The mapping at a national level resulted in a new attitude toward small-scale sewage. The emissions from these installations were compared to emissions from municipal treatment plants and off-grid sewage was pointed out as an important source of eutrophication of the Baltic Sea. The debate focused on the fact that nobody really knew how many small sewage facilities existed and that their technical status was unclear. The industrial organization, Swedish water, stated that the Swedish works had a high degree of purification of phosphorus and that it would result in unreasonable marginal cost of purifying more. The Swedish

Farmers' Organization (LRF) claimed that agriculture basically had done everything reasonable. Thus, in many cases, off-grid arrangements were at the center of attention when it came to combating eutrophication. The Swedish EPA initiated large projects to draw attention to small sewage and overfertilization and launched an inspection campaign called 'Small scale sewage – not a crappy topic' (Små avlopp – ingen skitsak) (Blomkvist, 2023a).

The Agency also published new general advice on small drains in 2006. It specified basic requirements and introduced the concepts of normal and high protection levels, for either environmental protection or health protection. For health protection, the previous attitude was maintained (the advice from 1987) that the sewage release must not mean an increased risk of the spread of infection and the protective distances were maintained. For environmental protection, recommended reduction rates were formulated close to the requirements postulated for large sewage treatment plants. The advice from 2006 recommended that a facility in the so-called normal level of protection should be able to remove 70% of total phosphorus during its lifetime. For small scale and off-grid arrangements, this level is considered quite high. Furthermore, as mentioned, it was also in 2006, when the new national water law was accepted, replacing the law of 1977 that environmental protection was firmly established as a criterion alongside individual health as a foundation for public engagement in water and sanitation. In fact, environmentally motivated regulations continued to put higher demands on both on- and off-grid sanitation. Requirements for the treatment of wastewater have increased from the mid-1970s following the heavy expansion of sewage plants between 1965 and 1975 and government subsidies for high-grade biological and chemical treatment. As mentioned, in an interesting article, Söderberg *et al.* (2022) investigate the history of alternative sewage solutions in Sweden since the 1980s. Resource recovery such as urine separation and usage of fertilizers from sewage sludge has been in focus. The authors conclude that although many promising attempts have been done, alternative sanitation solutions have met lots of resistance, mainly due to the inertia of the existing system.

To conclude, the articulation of publicness in off-grid water and sanitation was quite weak until environmental concerns and especially eutrophication of the Baltic Sea was put high on the agenda in the end of the 1990s. Since then the local arrangements, especially in sewage, has been a topic for much debate in municipalities around Sweden, concerning environmental demands put on off-grid sewage. I do not have the expertise to determine who is right and wrong, but there are actors claiming that municipalities place excessive demands on treatment in individual sewers and that the fear of eutrophication is exaggerated. The argument is that the soil retention in most cases takes care of the phosphorus content (Blomkvist, 2023a).

8.5 WATER AND SANITATION TODAY

Water provision and the sewerage system is presently regulated by the Public Water Act (SFS 2006:412 revised from 1 January 2023). It states that the municipalities have a monopoly in WS services and the responsibility for the whole *local water cycle*: from the water source via purification, distribution, use, wastewater treatment and return to recipients. However, it is important to point out that it is not a responsibility throughout the country, but only in those areas where 'for environmental or health reasons it needs to be arranged in a larger context' according to Section 6. In the 290 Swedish municipalities, there are around 1750 waterworks and just over 2000 municipal sewage treatment plants. In many cases, municipalities cooperate over borders in municipal and even regional organization managing water and sewerage. A majority of the 290 Swedish municipalities manage water and sanitation in a separate municipal administrative unit (61%). A minority 3% has a municipal WS company, 14% run multi-utility organizations where water and sanitation and other services such as electricity are managed jointly. About a quarter of the municipalities (22%) manage their infrastructural systems in cooperation with other municipalities (Bennich *et al.*, 2023; Blomkvist, 2023a).

The list below based on Christensen (2015) is not complete but gives a picture of the many stakeholders in water and sanitation:

- The municipal water and sewage department/company
- The municipal environmental division (overseer)
- The municipal planning and housing development division
- The municipal roads and streets division
- The County (Länsstyrelsen)
- The National Environmental Agency (Naturvårdsverket)
- The Ocean and Water Authority (Havs- och vattenmyndigheten)
- The Swedish Food Agency (Livsmedelsverket)
- The Public Health Agency (Folkhälsomyndigheten)
- The land surveyor (Lantmäteriet)
- Sweden's Geological Survey (SGU)
- Sweden's Municipalities and Counties (SKL)
- The Swedish Meteorological and Hydrological Institute (SMHI)
- Swedish Water (Svenskt Vatten)

The many stakeholders in water and sewage are also reflected in laws and regulations (a full list in Blomkvist, 2023a). As mentioned, we have the general Public Water Act (SFS 2006:412) as the juridical foundation which regulates larger sewage systems and their areas of operation; individual sewers can be found in the Environmental Code SFS 1998:808 and in the regulation on environmentally hazardous activities and health protection SFS 1998:899. Other areas are covered by the Planning and Building Act (PBL), SFS 1987:10 and the Construction Act, SFS 1973:1149, the Environmental Protection Act, the Health Protection Act, The EU Wastewater Directive (1991), the Environmental Code (1999) and the EU Drinking Water Directive (2020). Below is a simplified account of the division in responsibilities concerning sewage (Blomkvist, 2023a):

- The *Swedish EPA* is responsible for guidance on sewage plants that are sized from 200 person equivalents (pe) and up. In addition, EPA is responsible for guidance regarding dry toilet solutions and waste and cycle issues linked to waste related to 'small' sewage plants.

- The *Swedish Sea and Water Authority* provides guidance on 'small' sewage plants for domestic wastewater and are designed for up to and including 200 pe as well as 'small' sewage plants that are designed for up to 200 pe.
- The *County Administrative Board* is the supervisory authority for the large sewage treatment plants, that is treatment plants subject to a permit according to the environmental assessment regulation (2000 pe or more). Five county administrative boards are appointed by the government to be the water authority in each of their water districts. The water authorities are responsible authorities for management plans and action programs and for deciding on environmental quality standards.
- The *municipalities* are the supervisory authority for sewage plants designed for less than 2000 pe. As touched upon above, it must be noted that the municipality has two roles to play in the provision of WS services: supervision of health and environmental issues, and the traditional role of a system builder. These tasks are managed by separate organizations, one (the utility) deals with the physical building process and another (often the environmental division) supervises health and environmental issues.

It is easy to see that the regulations (and the organizational design) on water and sanitation, in the words of researcher and environmental lawyer Christensen (2003), are a 'patchwork.' There is no clear logic governing legislation and some rules from the middle of the twentieth century and earlier still are overlaid by recent EU regulations and directives. The multifaced and non-coherent appearance of water and sanitation is because water and sewage touch upon so many sectors in society. What is lacking is a state level system builder or controller responsible for the local water and sewage cycle. In my opinion, the *local* cycle cannot be left to the municipalities alone, especially not the small ones.

As mentioned, historically, the WS system is not as well aligned and cohesive as the road system. There is no central system builder at the state level and, again, WS systems do not exhibit a strong *vertical integration* (Blomkvist & Larsson, 2013; Blomkvist & Nilsson, 2017). Instead, we see a development toward *horizontal integration* where municipalities have created horizontal linkages and formed inter-municipal bodies (Alm & Paulsson, 2023; Alm *et al.*, 2021). We can also see what could be labeled a stronger vertical integration from an *institutional* point of view since the 1960s and up until the present, when the state takes a firmer grip of the sector by subsidizing the building of treatment plants and enforcing environmental legislation on both a national level and a European level.

doi: 10.2166/9781789063981_0097

Chapter 9

Carriers of technology and publicness in water and sanitation

When piped drinking water and later piped sewage were introduced in the larger cities roughly between 1860 and 1920, water and sanitation (WS) gradually became an integrated infrastructural system, although not yet with a national coverage. As in the road and street sector, we can see new technology being introduced by a new set of actors taking on the role of system builders or system promoters alongside the municipalities (not the state in this case). The new actors helped in changing the articulation of publicness toward health and sanitary issues and later toward environmental problems. As mentioned in Chapter 4 on roads and streets, I call these actors *carriers of technology and publicness* (Edquist & Edquist, 1979).

This chapter starts with a discussion on the new system builders promoting WS technology as a remedy for health issues and later as a solution for environmental problems, in the sector, followed by a discussion of the development of a certain systems culture in water and sanitation.

The sections on system builders and systems culture have some important limitations. I focus on the earlier history of water and sanitation and do not include a discussion on the new professional groups entering the sector when the environment took center stage. As mentioned, water and sanitation were gradually divided in two parts with overlapping logics and different scientific foundations. This meant that the municipalities were given two tasks in WS provision: the traditional role of system builders and supervisors of health, which were their responsibilities from the beginning, and environmental issues. Thus, from around the 1950s, and accelerating, environmental science and its practitioners became important carriers of technology and publicness in water and sanitation. It is my strong belief that this process is a much needed topic for future research.

9.1 SYSTEM BUILDERS AND TECHNICAL DEVELOPMENT

In water and sanitation, we do not see a clear central system builder role compared to roads and streets and the articulation of publicness was weaker in this sector. State and municipal authorities had the ambition and they surely tried to manage outer sanitation, but most often their efforts were fruitless. In water provision and inner sanitation, public involvement was even less apparent and came later. This historical heritage, I would argue, influenced the way modern water and sewage was introduced in the second half of the nineteenth century and onwards. Lacking a clear public agenda and a recognized state system builder or system promoter, the field was open for other groups. Medical doctors, helped by political opinion leaders, were the first group on the stage putting arguments from the Sanitary movement at the center of the debate. Already in 1851, when the first suggestion for a municipal system for piped water in Stockholm was put forward, the Swedish society of physicians started a campaign for modern water provision. The doctors' arguments came from a similar committee in Copenhagen, and most importantly from the sanitary reforms described in Chadwick's *Health of Towns Enquiry*. The arguments of the Society of physicians were later picked by Wilhelm Leijonancker in his design of the water supply system in Stockholm. The medical doctors were soon followed by municipal engineers and commercial consultants became interested.

Furthermore, Sweden did not have any real knowledge on WS-construction within its borders and technical know-how was, as been indicated, imported from Great Britain. In the second half of the nineteenth century though, 'in-house' expertise was built up and it was done mainly by three persons: Wilhelm Leijonancker, of course (1819–1883), Josef Gabriel Richert (1828–1895) and his son Johan Gustaf Richert (1857–1934). They were all involved in a majority of Swedish WS projects in the late 1800s, establishing domestic water and sewage system expertise. This led to a teaching position in water conduit, sewage, and waterpower at the Royal Institute of Technology in Stockholm from 1898, held by Johan G. Richert, who in 1903 was appointed the first professor in water construction. In fact, Sweden had no higher education for engineers until the mid-nineteenth century when the Technological Institute of Stockholm (Teknologiska institutet) was reorganized in 1846 and given resources for higher education. It was renamed the Royal Institute of Technology (KTH, Kungliga tekniska högskolan) in 1877. The Royal Corps of Civil Engineers (Kungliga Väg- och vattenbyggnadskåren) had been founded some years earlier, in 1851, and it consisted of the first Swedish engineers with higher technical education, either from the military academy at Marieberg or the Technological Institute. Their task was to assist in large public works such as canals and railroads, as well as to supervise the military engineering in times of war. As Sweden did not take part in any wars, the Corps was almost entirely dedicated to *civil* engineering in the true sense (Hallström, 2003; Tjulin, 2002).

Johan G. Richert was influential also in the technical consulting business. In 1902, he started the Water Building Bureau (VBB) which was to become Sweden's largest consultancy firm in municipal engineering and waterpower. Engineers as the three mentioned above, VBB and other technical consultants had an enormous influence in establishing a large-scale water and sanitation sector in Sweden already from the start but even more so in the big boom of WS systems after WWII (Söderholm, 2013). As mentioned, historian Karolina Wiell (2018) also underscores the importance of the municipal engineers and their successful strategy to transform sanitary issues from the medical to the technical domain.

In water and sanitation, the municipalities played a crucial role as system builders, especially after the municipal reform of 1862, and the organization that became most influential, and in fact can be seen as a system builder of sorts, was the Swedish Association of Municipal Engineers. As mentioned, they also functioned as carriers of new technology and publicness in modern street management and the association was founded in 1902 by high-ranking engineers from municipal administrations in Stockholm, Gothenburg and Gävle and very soon other municipal engineers from Sweden's larger towns joined the ranks. The association of municipal engineers worked in many areas to disseminate knowledge on system construction and management. Perhaps their most important contribution has been their successful testing and standardization of pipes, construction methods, tariffs, etc. and not the least their efforts to provide statistics on all sorts of issues related to water and sanitation. In 1962, water and sanitation related areas were transferred in the newly founded Swedish Water and Wastewater Works Association which today is called Svenskt Vatten (Swedish Water), an industry organization which represents most of Sweden's municipal water and sewage utilities (Tjulin, 2002). Swedish Water is still the most important provider of knowledge in technology, jurisprudence, and management. Their influence relies on the fact that many municipalities are small and not able to build up their own expertise in the area. They simply must trust Swedish Water. It is worth pointing out that water and sanitation is very different from other infrastructures in Sweden with its lack of a state level system builder. The special dynamics of interest groups, scientific and technical expertise, commercial interest and state and municipal strategies deserve their own doctoral thesis.

As touched upon above, pre-modern WS technology for a long time had a distinct local character using manual labor to fetch water from wells, rivers or lakes and disposing of waste and excrement in cesspools, privy vaults or into street gutters. In larger towns, so-called proto systems were tried but never really became a reality in the long run. Apart from institutional factors such as lack of organizational and financial capacity, the piped proto systems had a crucial technical drawback: the main pipes were made of wood and smaller service pipes were often made of lead. According to Rosen (2015), discussing the situation in England, in the beginning of the industrial revolution sometimes steam pumps and iron pipes were applied but it took a long time until iron became the dominant material. Water quality improvements also took a long time to develop. Purification technologies such as new methods to filter water with slow sand filters, originating in England around 1830, were gradually introduced. As already mentioned, the situation was similar in Swedish towns and when the piped water and later sewage systems were being introduced, most of the technical inspiration and know-how came from Great Britain.

The expansion of the Swedish sewage systems took place until about the 1950s with a mixture of combined systems (wastewater and storm water in one pipe), separated systems (wastewater in pipes and storm water in ditches) or duplicate systems (wastewater and storm water in two pipes). According to statistics in 1942, 74% of 132 communities applied only the combined system. The combined system was preferred up until the 1950s. Today, the general sewage network consists of approximately 13% of combined pipes. But an extensive private service pipeline network is also connected to the public sewer network. The private drinking water network and wastewater system make up just over 20% and the private storm water network a full 78% calculated as km of pipeline.

Approaches to stormwater management gradually changed from the 1970s with the introduction of ideas of 'Long-term sustainable stormwater management:'

- Until 1975, the focus was only quantity problems and diversion to the nearest recipient.
- Between 1975 and 1995, attention was drawn to the fact that stormwater contained pollutants that needed to be taken care of.
- From 1995, the term 'Gestaltning' was added with storm water as part of general urban planning.

Thus, in the Climate and Vulnerability Investigation (SOU 2007:60), the perspective for stormwater management was lifted from a 'simple pipe issue' to an issue for 'community planning,' The realization was that pipe systems cannot fully handle extreme torrential events, so the diversion must take place in so-called downpour roads. Today, this is a generally accepted insight and there are many examples both in Sweden and abroad. Superficial downpour management is a prerequisite for being able to handle extreme weather. But there are problems to implement this holistic strategy, not the least because no central actor has full control over stormwater management (Blomkvist *et al.*, 2023).

9.2 SYSTEMS CULTURE

Building on the above-described movement toward large and centralized piped water and sewage facilities, Cettner *et al.* (2012) claim that problems in WS systems today and especially concerning combined sewers for both excrement and stormwater can be explained by institutional and technical path dependence and by a systems culture based on a 'pipe-bound mentality' of actors within water and sanitation defining water and sewage as a centralized and piped underground infrasystem.

As mentioned, from the 1870s, towns applied the logic of the old gutter-based sewers when building new underground pipes, because sewerage had always been meant to drain stormwater and to get rid of excess wastewater. Thus, *combined* pipes (in contrast to separate pipes) became the most common design choice. Perhaps involuntary, Sweden's environmental protection initiatives in the 1960–1970s contributed to system inertia because public funding was mainly directed toward sewerage.

Cettener *et al.* argue, building on Drangert and Löwgren (2005), that the extensive public funding and the many institutions, and organizations dependent on the existing system created a 'self-sustaining systems culture' among WS actors which streamlined the design, operation, expansion, and management of the system. The system culture was strengthened by the common educational background of the actors. The infrastructural systems were relatively open and flexible concerning design choices in the initial stages of construction at the end of the nineteenth century. But quite soon, centralized and large-scale water and sanitation became increasingly more rigid and difficult to change, partly due to the large investments (sunk costs), the durability of the technical components in the system and as mentioned because of the establishment of a supporting social and political environment. All of this consolidated the centralized system as the Swedish water and sewage strategy and contributed to system inertia, an inertia that was strengthened by the system culture of water and sanitation.

When new ideas of local management and disposal of stormwater came into the debate in the 1970s, the systems culture became obvious. The greatest challenges for decentralized storm water management lay in a conservative systems culture among stakeholders, which effectively slowed down the implementation of these ideas. It seems as if Cettner *et al.* (2012) have a strong point when they claim that the historical choice of tying stormwater management to a pipe-bound centralized water and sewer system has made sustainable solutions almost impossible: 'To achieve a more sustainable stormwater management program, municipalities must break away, both physically and mentally, from the traditional system and its culture.'

Furthermore, the storm water issue was another institutional factor because streets and water and sanitation were managed by different organizations. In the 1940–1950s, municipalities tried to move away from the combined sewage system to some extent financed by the state and the Royal Roads and Waterworks Agency. However, the state subsidy was targeted at the drainage of municipal streets by using separate storm water pipes, but the grants did not include the associated service lines from municipal or private housing. The result was that hardened surfaces, such as rooftops, remained on the old combined sewer pipe. It is an interesting question if the design of the state subsidies can explain the fact that this so-called 'additional water' still burdens the municipal treatment plants. Water engineer Hans Bäckman at Swedish Water claims that it was unfortunate that the drainage of the street and the installation of stormwater wells ended up on the 'street side,' while the rest of the stormwater was diverted to the 'WS side' and its combined stormwater pipes. To some extent, this divide between street drainage and storm water still exists today (Blomkvist *et al.*, 2023).

As will be elaborated in section 10.5, right now many municipalities make efforts to remedy the problems created by combined sewers. To meet challenges from global warming and more rainfall, it is not considered possible to rebuild or re-size the sewage pipes which of course would be very expensive. Instead, the whole municipal planning process must be revised toward a more holistic and preventive approach (what Cettner *et al.* call 'a nature-oriented and local solution'). For example, one must consider how hard surfaces such as roof tops and streets affect storm water management so that effects are reduced. To deal with these issues, departments dealing with building permits, environment, etc. must be included and the water utility or the street department will not any longer be the only municipal bodies responsible for storm water. In a new addition to the national water law from 2023, this ambition is strengthened in Section 6b which stipulates that every municipality must have a 'Water Service Plan' addressing these issues. This means that the new Section 6b tries to fix what Cettner *et al.* (2012) argue is the main problem by encouraging local disposal of stormwater.

doi: 10.2166/9781789063981_0103

Chapter 10

Comparing publicness in municipal infrastructure

In this last chapter, I sum up the most important traits in the history of streets, water and sanitation and focus on comparing these three areas in relation to how the articulation of publicness has varied between them. To close, there are a few sections discussing some general conclusions and insights from the research regarding future challenges in municipal infrastructure management.

10.1 PRE-MODERN ARRANGEMENTS BEFORE 1800

Roads and streets have, as indicated, been articulated as a public domain for a long time. The Swedish state and the towns and municipalities have strived to create laws and institutions promoting road and street building and maintenance, at least from the Middle Ages. I would argue that the road and street network was a public concern long before it became an infrastructural *system* in our modern meaning and that the authorities showed a clear public ambition long before they really became system builders.

The publicness of roads and streets was for a long time associated with the so-called principle of 'utility and interest,' meaning that landowners, in towns and in the countryside, living in proximity were to manage the road or street because they were seen as the beneficiaries of the service delivered. However, the strong tradition of local self-governance meant that street and road management was exercised with a gentle hand and that the goal was to reach amicable agreements. Without some 'communalism' and the goodwill of the farmers and property owners, no roads or streets were built or maintained.

In the first Swedish national legislations on water, the private, non-public, character of drinking water is visible. Publicness in water was not articulated as drinking water provision in the pre-modern era. Instead the national laws on water focused on *defensive* projects such as drainage and *lucrative* project such as waterpower. Thus, contrary to the situation in ancient Rome, in Swedish legislation drinking water was generally treated as a private good and a private affair (Res private) and a productive resource rather than a public good and public affair (Res publica).

Furthermore, there is no sign in the older history of rural areas, such as village ordinances or by-laws, legislation, or court proceedings, that treats drinking water as a public concern. Drinking water provision in the countryside of Sweden was simply not regarded as a common or public responsibility, at least not in any formal sense in written laws and regulations.

In towns however, due to higher population density, the situation was different. Through the seventeenth and eighteenth centuries in Europe, water provision in towns in many cases partly relied on *proto systems*. In the larger Swedish towns like Stockholm, Gothenburg and Malmö, attempts were made using various types of proto systems for piped water during the Middle Ages and the following centuries. Some of these proto systems can perhaps be seen as something in between older forms of service arrangements and modern infrastructure, the *missing link* in the evolution of municipal infrastructure.

Common wells for water supply certainly existed but drinking water was not the only, or even the main, reason for building and maintaining them. Rather, public drinking water could be seen as a welcome side effect of the basic motives for the common wells: fire protection and street cleaning. Thus, in the articulation of publicness concerning water, fire security and tidiness were articulated more strongly than provisioning of drinking water for the people. It is my belief that there is an exaggerated and unfounded notion of the degree of publicness in drinking water in historic times due to an interpretation of history from a present-day standpoint. Today it is almost impossible to think of a situation where water provision is not a central public concern in Sweden, which makes it easy to believe that this has always been the case.

Concerning sanitation, the situation was somewhat different. In the pre-modern period, public sanitation relied on open diches and gutters to deal with street cleaning and storm water. From around 1800, sanitation was gradually associated with human health and wellbeing, a matter that increasingly became a public concern. This movement changed the motives in outer sanitation, toward individual health. It also gradually transformed inner sanitation into a public concern and responsibility.

In conclusion, it is important to note that many ancient civilizations and towns had WS arrangements that were much more advanced than in Europe before the nineteenth century. In fact, one of the most surprising traits in European history is the weak public involvement from the fall of Rome and up to the modern era. Even more surprising is perhaps, at least in Sweden, the low public engagement in drinking water in the pre-modern period.

10.2 SYSTEMIZATION AND INFRASTRUCTURE BUILDING 1800–1920

The state's interest in public roads during the Middle Ages and up to the nineteenth century was essentially based on the central power's need to control the territory, mainly for military reasons, which was gradually complemented with a desire to promote trade and manufacture, both in the nation and in towns. Industrialism affected the sector and road and street keeping was articulated in new ways. First, agriculture became more market-oriented and the need for transport outside the absolute immediate area increased. Second, industries grew that required roads for its raw materials and products. The farmers who were obliged to maintain the roads came to see the road burden as unfair because the new industries did not have to pay for the roads they used. Third, the building of the railway network meant that the need for roads increased as goods and passengers had to travel to and from the stations. Fourth, the route of the railway meant that population and business were concentrated in connection with the nodes of the railway network. This development can be characterized as the 'industrialization of roads.' Moreover, adding to these general trends in the industrialization process affecting the articulation of publicness, technical advances were of course pivotal for the development of these areas into infrastructural systems. Industrialization also meant that more people moved into towns and industry's need for better transport facilities affected road and street construction, pavements, but the real challenge to streets came with the automobile.

The most important contextual factor influencing the articulation of publicness in roads and streets up until the first half of the twentieth century was undoubtedly the strong tradition of municipal independence and self-governance. The history of roads and streets shows a high level of local independence and negotiating power in relation to the state and municipal authorities. It is quite clear that towns had a high level of freedom in dealing with street and road issues and some cities had started to take over road and street maintenance from the middle of the nineteenth century and financed this with taxation. Furthermore, as mentioned, in the larger towns such as Gothenburg and Stockholm, the magistrates or the Burghers decided to manage both roads and streets as a public undertaking even earlier. The Municipal Act of 1862 gradually changed both road and street keeping in the cities, toward more public engagement and it seems probable that most towns had taken over both street and road management as public undertakings in the beginning of the 1920s.

But municipal independence also slowed down a reform in the road sector. It was difficult to reform road maintenance due to far-reaching individual and local self-determination, where many changes could be blocked by individuals or by local communities. The nationalization of the road network in 1944 was a drastic move to completely disconnect road maintenance from local interests. Nevertheless, the conservatism found in public road keeping in smaller municipalities on the countryside was not present in larger cities. The old habits of street and road management faded away faster in the towns, and they were able to modernize the sector earlier on. In the towns, the strong municipal independence did not cripple road and street keeping as it did in the countryside. The city authorities used their strong self-governance to adapt the sector to modern demands.

Water and sanitation became gradually more public from the first half of the nineteenth century. This process impelled authorities to take on a more active role and off-grid service *arrangements* for water and sanitation were slowly transformed into on-grid *infrastructural systems* at a time when the municipalities were able to shoulder the system builder role.

The first suggestion for a municipal system for piped water in Sweden was put forward in 1851 in Stockholm and motivated by hygienic, social, medical, and economic advantages of piped water, not to mention its importance for firefighting. Wilhelm Leijonancker was commissioned to design the water supply system. I believe that if one wants to find a 'Swedish model' of sorts for WS infrastructure, Leijonancker's design plan is a good starting point. It certainly influenced many of the following towns building water, and later sewage systems in the decades after Stockholm.

I also want to stress the profound influence of Chadwick's *Health of Towns Enquiry*. Leijonancker openly referred to the Sanitary movement and it reports on the health situation in England, comparing and adjusting the results to Swedish a context. Most of his general motivations for the piped water systems are blueprinted on Chadwick and the Sanitary movement.

The initiative to build a piped water network in Stockholm was launched 10 years before the acceptance of the Municipal Act of 1862 and the system was ready for operation 10 years before the Public Health Act of 1874. As mentioned, many other towns followed suit and Sweden already had functioning water works in approximately 10 towns by 1874. Also in the 1870s, piped sewage was introduced and motivated by the same health reasons and because the piped water systems brought much more water into the towns that had to be disposed of. This development strongly indicates that the contextual factors discussed had a profound influence on the articulation of publicness and the development of water and sanitation and that the formal legislations were codifications of a more positive attitude in society toward publicness and interventions from the state and from municipalities in domains previously seen as private.

10.3 MATURING INFRASYSTEMS 1920–1980

Eventually in the 1920–1930s, economic efficiency became more important and central control was strengthened in 1934 when yet a new Road Act was introduced; the most important change was the formal abolition of road keeping in kind. Something that de facto had already happened. In 1944, the public roads in Sweden were nationalized and the utility and interest principle moved up to a societal level. Now publicness was articulated as the entire kingdom's joint obligation to keep roads and streets. Also, from the 1930s and especially after WWII, publicness in the road and street sector was foremost articulated through the automobile. The national road plan for Sweden in 1958 was an ambitious effort to remodel the public road network to fit the demands of mass motorization. In towns, the municipal engineers worked hard to adapt streets and public places, and the whole town, to motorized traffic, especially the private automobile. A spectacular effort of public engagement was the decision to switch to right-hand driving in 1967. Towns, municipalities, and the whole country showed an unprecedented will to embrace publicness in the traffic sector.

During the interwar years, more and more towns built piped WS systems and after WWII this development also reached the countryside. The motive was to include rural areas in the Swedish welfare society and an important feature was a widespread expansion of water and sanitation starting in the early 1930s. Water and sewage systems grew fast in Sweden during the first half of the twentieth century. The growth was of course coupled with increased numbers of users and also because of increased public investments and the forming of institutions and organizations supportive of the system, all contributing to system inertia.

To sum up, there were ideological, environmentalist and economic motives for municipalities to push for centralized solutions. With the help of government subsidies and optimistic assessments of population growth, large, expensive sewage treatment plants had been built. The investment made in a central solution had a decisive influence on the choice of subsequent investments.

The account given so far might give the impression that the WS infrasystem is totally dominant, comprehensive, and covering all parts of the country. And it is certainly true that we can see a development from privately managed WS arrangements to public municipal infrastructure, the articulation of publicness is loud and clear. Nevertheless, it is very important to note that large parts of the countryside did not get access to piped water until way up into the 1970s and that Sweden still has a large part of its population still depending on private arrangements. These, what might be called *pre-modern* WS arrangements, are still to a high degree present and consist of private property owners arranging for their own off-grid solutions.

The private character is still a reality and public authorities accept these pre-modern styled service arrangements, especially concerning drinking water from private wells where regulations on water quality certainly exists but they are not as forceful as rules for private sanitation/sewage arrangements. In fact, the public control of private wells is almost non-existent according to informants. The reason for the difference in attention and the view that sanitation is a public concern more so than drinking water, that water quality is up to the individual and still not fully articulated as a public responsibility, is because private sanitation arrangements have negative externalities. Private sewage solutions can possibly contaminate ground water sources, that is the well of your neighbor, and at the same time pollute nearby water courses and recipients. According to the environmental law, the owner of a sewage facility has an obligation for proper maintenance to protect the environment. Today, urbanization, peri-urbanization, and global warming push the WS system to develop decentralized and hybrid solutions. But changing a piped system is difficult. Most often infrastructural systems develop gradually through incremental innovations.

To conclude, in the 1980s infrastructure for roads and streets and water and sanitation had reached maturity and we have not seen any larger expansion of the infrasystems since then. Both sectors were firmly established as public responsibilities; the articulation of publicness were (since quite a while) completed. The two most important factors affecting municipal infrastructure during the last decades have been the combined realization that we have a large maintenance deficit, which were discussed in the beginning of the book, and at the same time a heightened awareness of the challenges posed by global warming (more on this below).

10.4 OWNERSHIP AND FINANCING

A central component in the articulation of publicness in streets, water and sanitation has of course been ownership of the service arrangements or infrasystems and how the building and operations has been financed.

In public roads on the countryside, the state has had a strong interest in articulating publicness since at least the Middle Ages, but it is not correct to talk about ownership in any real sense until the first half of the twentieth century. For many hundred years, the state had to rely on cooperation and goodwill of the farmers and the county authorities to manage the road sector based on the principle of 'utility and interest' of the property owners living nearby. It must be noted that in civic roads, the involvement of property owners and the principle of utility and interest are still a living fundament.

Real state ownership of public roads came gradually, accelerating in the 1920s when the principle of interest and utility and road keeping in kind by farmers and property owners became obsolete, in the 1930s when the principle of 'right of way' was legally established and finally, in 1944 when public roads were nationalized.

In municipal streets, an area for a long time only concerning larger towns, the city ownership was articulated a bit stronger. Street keeping in towns was based on the same principles of 'interest and utility' that governed public roads. Nevertheless, streets were clearly, and since a very long time, articulated as a public, municipal undertaking and city authorities had the responsibility for public places and squares as well as public roads passing through the town. Furthermore, many of the larger towns also transferred street management under municipal/city authority already in the first half of the nineteenth century, long before public roads came under state ownership.

From an infrastructural systems perspective, it is quite clear that state and municipal public interests and ownership evolved gradually as other contextual factors in society changed the playing field of the sector. Nevertheless, I would argue that state and municipal involvement was very strong a long time before roads and streets were turned into infrasystems. Publicness came ahead of systemization.

Setting these differences in ownership aside, roads and streets have always been financed through taxation. Early in history, the tax was paid in kind for maintenance and as a general tax for road and street building. There are very few examples of road or street pricing using fees or tolls in Sweden. The financing by taxation is a clear sign of a strong articulation of publicness in roads and streets that make this sector differ from water and sanitation which was more often financed by individual fees as will be discussed below.

Drinking water provision in the countryside had a weak articulation of publicness for a very long time. Instead it seems that informal rules and customs on water sharing as a moral obligation governed provisioning of drinking water.

In towns, drinking water had a somewhat stronger articulation of publicness. The city authorities made some efforts, in some towns, to build proto systems for piped distribution and common wells were certainly quite common. However, drinking water distribution was not the prime motive for these service arrangements. Instead firefighting and street cleaning were the areas where publicness was most strongly articulated. Drinking water was a welcome side effect.

Thus, public ownership of early service arrangements for drinking water was not really an issue. Municipal ownership first became a reality from the beginning of the nineteenth century when many contextual factors in society reinforced the growing public interest in drinking water. These contextual factors became apparent in the Sanitary movement and in the National Health Act of 1874. But already in the middle of the century, Stockholm and other large towns stared to discuss and plan for municipal water works and a piped distribution network.

Following Wilhelm Leijonanker's suggestions, the norm (the 'Swedish model') of municipal ownership was firmly established in most towns and later in almost every municipality. Thus water works were considered public in stronger sense, including a demand on public ownership, in contrast to the earlier gasworks, that often were run by private companies in the beginning. In this respect the piped water system, although local and under the municipality, was more like the state-owned national railroad network planned and built in the same period. Public ownership has lasted until this day and age and water and sanitation is acknowledged as a *natural monopoly* and therefore not an area for private ownership and commercial interests. It is fair to say that in drinking water publicness came, not before as in roads and streets but in conjunction with systemization.

Historically, the articulation of publicness has always been stronger in sanitation compared to drinking water but weaker than in roads and streets. Outer sanitation, street cleaning, storm water disposal in gutters and waste removal were closely linked to street keeping in Swedish towns and therefore a more obvious public domain. From around 1800, both outer and inner sanitation were gradually associated with human health and wellbeing and increasingly became a public concern. Public interest now focused on earlier private practices such as house cleaning and personal hygiene and Swedish towns took a firmer grip on excrement handling by a gradual takeover of latrine barrel collection by publicly employed personnel. However, sanitation defined as an infrasystem was directly linked to the establishment of modern piped water systems. Piped sewerage just like piped water distribution also came under municipal ownership in most towns and municipalities. In fact, sewerage was considered a clear natural monopoly, even more so than piped drinking water, which made municipal ownership the obvious choice.

Regarding the financing of water and sanitation, the most common choice has been different forms of user fees. It is true that some towns distributed piped water 'for free,' that is paid over the municipal budget based on taxation, but this model for financing did not last for long. Sanitation, both inner and outer, was mostly financed by individual fees and in some periods as work in kind by the property owners. As mentioned, the differences in financing between these areas of service arrangements give a clear indication that roads and streets had a stronger articulation of publicness compared to water and sanitation. It is interesting to note that the reliance on individual fees prevailed when the older service arrangements for water and sanitation turned into infrastructural systems and that fees for every connected user to the municipal grid still is the chosen business model all over Sweden. Ownership and financing of infrastructural systems truly show the dependence of historical choices made and the importance of understanding the historical context that shaped the transformation of off-grid service arrangements into gridded infrasystems. The institutional path dependence and the historical heritage are apparent in the evolution of municipal streets, water and sanitation.

Following from the above, I want to stress a fundamental difference between the road and street sector compared to water and sanitation from an historical perspective. In roads and streets, there has been little or no debate whether they ought to be a public responsibility or not. Publicness, in this general sense, has been clearly articulated for a very long time. Instead, the debate has centered on the question of who, which groups in society, ought to be responsible to carry out the tasks in road keeping. Who must carry the road burden? Thus, roads and street management were not so much affected by the changed attitudes discussed above related to the possibilities for the municipalities (and the state) to intervene in earlier private matters. The road and street sector had 'always' been articulated as a public domain and it was not until the end of the 1920s that public bodies took over the practical tasks in road and street keeping.

In water and sanitation however, for a long time, the debate was much more heated on whether these areas should even be seen as public in the first place. The articulation of publicness in water and sanitation was related to the specific contextual factors discussed above leading to the Sanitary movement and operationalized the Health Act of 1874. These factors started to influence water and sanitation in the beginning of the nineteenth century and were of course a part of the changed relationship between public and individual spheres of society, which gave a mandate for the municipalities, to intervene in various areas. The articulation of publicness in water and sanitation was much more closely linked to the question on which groups that were supposed to perform the tasks at hand and public responsibility and ownership were closely related to this articulation process.

To sum up: for roads and streets, publicness was articulated long before actual public responsibility became a reality. In water and sanitation, the articulation of publicness was simultaneous to the public taking over the tasks. In other words, for roads and streets, the articulation of publicness predated the creation of an infrastructural system. In water and sanitation, on the other hand, the articulation of publicness went hand in hand with modern infrastructure.

10.5 MUNICIPAL INFRASTRUCTURE AND GLOBAL WARMING

As mentioned, global warming will affect society and municipal infrastructure. We will see both drought and more rain. Climate models show *flips* from severe drought to heavy downpours, so-called *compound extreme events* will become more frequent which will lead to increased risks in 'health, ecosystems, infrastructure, supply and food,' according to the The UN Intergovernmental Panel on Climate Change (IPCC's) latest major report (Blomkvist, 2023a). Consequently, climate change effects force municipal utilities to implement change in many infrastructural systems that historically have been stable in the last 50–100 years. According to the United Nations there is '… a need to increase the resilience in traditional large-scale infrastructural systems …' and the UN call for 'bold and transformative steps which are urgently needed to shift the world on to a sustainable and resilient path' (UN Agenda 2030). The Swedish Environmental

Agency (Naturvårdsverket, 2016) warned that both water shortages and flooding will impose challenges for conventional technologies for water distribution and treatment. The world economic forum also identifies water scarcity as serious risk.

Thus, global warming challenges municipal infrastructure in both streets and water and sanitation. The common dominator and the most profiled risk area is storm water management. Furthermore, the biggest obstacles seem to be the many stakeholders in the infrastructural sector and that nobody really knows who is responsible for storm water issues (lack of rådighet). The responsibilities are often organized in silos where the actors deal with their own area with little or no coordination. Until recently, for example, the traffic administration in Stockholm dealt with storm water but in 2021, Stockholm water was given the responsibility for storm water management. This situation also affects the ability for innovation and future foresights, which seems to be left to outside consultancy firms. Therefore, more cooperation between actors in municipal infrastructure is needed to deal with the strain on both municipal streets and water and sanitation. Earlier management practices where these two areas were kept separate are no longer working. In the two biggest towns in Sweden, Gothenburg and Stockholm, cooperation and cross-sectoral work are emphasized and in Stockholm this need for action across professional and political borders is described like this (Blomkvist, 2023a):

A denser city is a challenge when available areas for handling stormwater decrease. At the same time, there is a need to provide space for stormwater because higher demands are placed on improved recipient quality. The buildings also need to be adapted to meet the effects of climate change as well as increased expectations to meet the needs of urban greenery. A traditional urban environment largely consists of hard surfaces. Here the natural drainage channels, which provide delay and infiltration, are largely replaced by technical stormwater systems. These changes give stormwater a very rapid runoff. The rapid runoff results in a reduced fixation of pollution. This means that pollution is instead added and burdened receiving lakes and streams.

A strong movement toward circular economy can be found in future oriented discussions on public infrastructure, especially in water and sanitation. The articulation of publicness is framed in a movement from waste disposal to reuse of resources and the slogan is: 'Today's treatment plant – tomorrow's resource plant' (Blomkvist, 2023a). In Helsingborg, in the south of Sweden, the municipality has started an acclaimed and often mentioned project called 'RecoLab – Pilot Recovery Plant for Sustainable Management of Wastewater and Food Waste.' From a historic perspective, the circle is closed, and we are back in pre-modern sanitation practices where latrine was seen as valuable resource and fertilizer in agriculture. As touched upon, this holistic view was also an early fundament for Edwin Chadwick and the Sanitary movement in the first half of the nineteenth century. The piped sewage system with its connected water closets were meant to transport fecal matter and waste to farmlands outside the city where farmers would pay for them and use it as fertilizers. This business model would finance the sanitary improvements in the cities, and at the same time benefit agriculture (Hallström, 2003). Using the words of Chadwick (see earlier), in Helsingborg they are trying to complete the circle and bring the serpent's tail into its mouth once again by moving from reuse to 'getting rid of' (kvittblivning) and back again to reuse. In popular terms, the 'three pipes out' project is presented like this on the web site: 'The new city district of Oceanhamnen in Helsingborg has created a solution for separating and recovering different kinds of wastewater and food waste at source.' Gothenburg also had a black water plant in 2007–2008 that was tested at full scale in an area with gray water separation (Blomkvist, 2023a).

However, and as mentioned earlier, to discover reuse and resource-oriented solutions in sanitation, one does not have to go so far back in history to the era of Chadwick. Sweden was in fact an early forerunner in this field. The pioneering attempts failed though, and the reasons were mainly connected to the inertia of the existing system (Vidal, 2022).

Nevertheless, technical and organizational innovation in circular economy is not the only strategy. Currently, the dominating trend to mitigate climate change effects and to meet expected population increase in Stockholm (and in London) is to build even larger sewage facilities.

London is constructing a 'Super sewer' to extend the old Victorian sewage system designed by Bazalgette. Thames water, the public WS organization, has hired Tideway (Bazalgette Tunnel Limited) to finance, build, maintain, and operate the Thames Tideway Tunnel under the HM Treasury's National Infrastructure Plan. Preparatory work began in 2015 and construction is now underway. The new tunnel system will be 25 km long and 7.2 m in diameter, will cost £4.3 billion to complete and is anticipated to be completed in 2025. This large-scale project is motivated by the prospect of increased rainfall due to climate change and an expected increase in population from 9 million to 16 million in 2160 (the Bazalgette system was designed for 4 million people). But the sewage produced by the increasing population does not seem to be the main problem; storm water overflowing into the river is the real challenge:

> Sewage is the last significant source of pollution in the river. The Super Sewer is the solution … The Thames Tideway Tunnel will protect the river for at least the next 100 years. (Blomkvist, 2023a)

Stockholm is also building a similar super sewer, and even if that name is not used the utility claims they are building 'One of the world's most modern treatment plants.' An extensive modernization of the plant in Henriksdal, south of the city center, began in 2015. The project also includes a 14 km tunnel stretching under the town from an earlier treatment plant in the west suburbs which will be closed. Construction of the tunnel began in January 2020, and it will be operational in 2026. The tunnel will also alleviate problems due to the many combined sewers in Stockholm (as mentioned, most sewers built before the 1950s in Sweden were combined).

Henriksdal's treatment plant is being upgraded and optimized with new biological treatment and one of the world's largest facilities with membrane filtration technology and the expanded treatment plant will be well equipped for future requirements (Blomkvist, 2023a).

Apart from the large-scale examples above, there is also a trend toward small scale and local alternatives in water and sanitation that exist alongside the large-scale facilities. As mentioned, even though Sweden has a well-developed WS system with around 1.5 million properties connected to municipal water and sewerage facilities, in 2015 almost 1 million properties, of which 500,000 leisure properties were not connected.

One example of the growing interest and actuality of small-scale WS can be found in Värmdö, a municipality outside Stockholm with 45,000 permanent inhabitants, which grows to around 100,000 in the summer. It is a coastal region with more than 10,000 islands and about 15,000 holiday homes. Today Värmdö has between 15 and 20,000 individual water and sewerage facilities in holiday houses, which often lack capacity for permanent living. Urbanization makes people increasingly relocate to the cities, but due to an overheated housing market, many move outside the city center to find affordable dwellings. This means that areas earlier used for summerhouses and recreation are exploited, putting pressure on infrastructural systems not well adapted for permanent living. During the last decade, around 3000 holiday homes have been transformed into permanent dwellings. In total, around 27,000 people live in houses not connected to the municipal grid and the utility in Värmdö has no knowledge of what sort of technologies are used by these private property owners. There are an increasing number of technical solutions, from private mini treatment plants to septic tanks, and decisions on technology taken by individual property owners at a free market. For the utility, the influx of people living permanently in refurbished summerhouses has changed the expectations on WS services and few, if any, accept dry solutions for latrine. Instead, there is a strong trend toward wet solutions (WC). As more areas are being transformed to permanent dwellings, Värmdö municipality foresees water shortages and serious sewerage problems in some areas. Furthermore, the utility predicts that effects related to the Corona pandemic are pushing for a trend where even more people settle permanently in Värmdö using digital communication technology to work from their homes. In 2020–2021, applications for building permits almost doubled (Blomkvist, 2023a).

The situation in Värmdö is not unusual in Sweden. Many municipalities close to larger cities experience simultaneous trends of urbanization and peri-urbanization that together with global warming push the WS system develop decentralized and hybrid solutions. And these trends also affect municipal street and road keeping. As discussed elsewhere, new inhabitants moving in from the city centers are used to a higher standard of street and road quality. They expect asphalt pavements, street lighting, not to speak of other facilities such as daycare and schools, and, perhaps most importantly, they demand that these services should be taken care of by the municipality. What we see is a clash of city dweller expectations and the reality in peri-urban municipalities. The articulation of publicness in municipal infrastructure has not stopped evolving and it still causes controversy (Blomkvist, 2010).

10.6 PUBLICNESS AND MUNICIPAL CAPABILITIES

The future challenges for municipal infrastructure are not only about dealing with global warming and its subsequent catastrophic events such as flooding and heavy rain. Just as important is to manage the maintenance deficit mentioned in the introduction. For both areas, the issue of municipal *capacity* is at the center. The capacity problems are related to size. For obvious reasons, larger municipalities and towns have the resources to hire experts and develop critical capabilities in their infrastructure departments which seldom can be matched by smaller units. Nevertheless, small, and even *shrinking*, municipalities still must deal with the same problems and follow the same regulations in, for example, environmental legislation (Syssner & Jonsson, 2020).

The problems with infrastructure maintenance are mainly due to the enormous need for re-investments because the networks have become outdated and fallen into despair: 'In short, the long period of underinvestment in maintenance and repair of critical infrastructure networks, often caused by short-term financial considerations and a run-to-failure mentality, has led to the growth of a so-called maintenance debt' (Alm *et al.*, 2021).

Alm *et al.* (2021) discuss different forms of municipal capacity that is needed to successfully manage streets (roads) and water and sanitation, and they investigate this local capacity in five interrelated aspects:

(1)   Technical capacity
(2)   Financial capacity
(3)   Institutional capacity
(4)   Political capacity
(5)   Social capacity

While the importance of each of the aspects will vary from municipality to municipality, they are interrelated and influence each other and will all to some extent influence the overall capacity of an organization to achieve its objectives. The difference between municipal street keeping and water and sanitation is described like this: 'For the road networks, maintenance is generally outsourced to contractors and there is also a large degree of tolerance for various standards on different road segments within and between the municipalities. Less used road segments are not as prioritized as those with heavy traffic. For the water and sewage systems, in-house technical capacity is needed as differences in water quality are not tolerated. Economies of scale mean that in-house capacity is translated into the creation of inter-municipal bodies.'

Leaving the differences between maintenance of streets and water and sanitation aside, it is perfectly clear that population size, economies of scale and the ability to recruit capable personnel are crucial for the municipal capacity expressed in the five aspects above. To put it bluntly, it is often impossible for smaller municipalities to present the technical, financial, institutional, political, and social capacity needed to successfully manage infrastructure.

Thus, the historical legacy of a municipal organization of these infrasystems has a distinct influence even today. For street keeping, the state level is influential, but the heritage of municipal self-governance has kept much of the area under local authorities. Concerning water and sanitation, the problems for small municipalities are even worse. They are under an even stronger culture of local independence due to the historical development of the area. In both cases, the legacy of local management makes it difficult to deal with the present-day problems related above. Municipal self-governance of streets, water and sanitation seemed like a natural choice some 150 years ago. Today, local level management creates tensions and difficulties not foreseen by the historical actors. Again, I argue that local management might well be a root cause of problems in water and sanitation today. What is lacking is a state level system builder or controller responsible for the 'local water and sewage cycle.' Although *local* in its character, WS systems cannot be left entirely to the municipalities, especially not the small ones.

In a general sense, history affects infrasystems in two ways. Firstly, in what I would like to call 'history ex-ante' (before the event), that technical and institutional choices made when designing infrastructure affects operations for decades, even centuries, ahead. Secondly, history affects infrastructure management 'ex-post' (after the event): in existing infrastructure the historical legacy, because of inertia and path dependence, limits the options to update the infrasystem to new contextual factors such as global warming. Katko *et al.* (2009, 2006) deliver a hard critique of the lack of strategic and long-term planning in water and sanitation infrastructure relating to the 'ex-ante' aspect of history. They show that strategic thinking and long-term planning was often missing in the early stages of infrastructure building. Instead, utilities and system builders often concentrated on short-term operative or 'opportunistic' management of services, ignoring the fact that the '…lifespan of these systems is very long – some parts serve more than a century which makes it necessary to introduce longer-term strategic and visionary thinking'. Furthermore, the authors also discuss problems with a lack of understanding of the 'ex-post' aspect of history and they relate cases where decision-making is based on current technical and economic conditions without addressing other issues such as the prevailing policy and institutional set up, as well as social and environmental considerations: 'Even more serious is the finding that, instead of identifying and assessing several options in the early phase of projects, too often only one option is considered by leaders, politicians or other related parties.' Katko *et al.* argue, and I agree, that although management of water and sanitation is inherently local, management paradigms should be extended to longer-term futures, including an awareness of factors hampering innovation related to historical legacy and path dependencies (Katko *et al.*, 2009).

10.7 ELABORATING THE 'PIPED PARADIGM'

Earlier I have discussed the *systems culture* in water and sanitation developing in connection to the massive expansion after WWII of piped WS and the building of large treatment plants all over Sweden in the 1960–1970s. This systems culture is often described as a 'piped paradigm' (Braadbaart, 2009) with its actors sharing a 'piped mentality.' The strong preference of expanding infrastructure by underground water and (gravity flow) sewage pipes has created a strong inertia and technical path dependence of the traditional system. The municipal utilities have a powerful position in the piped paradigm as they decide on where and when to expand and what type of technology is to be used in this expansion.

This description is quite accurate, but it needs to be nuanced and elaborated. The 'technological power' of the utilities is not the only explanation for the inertia in the existing system. It is not only the piped paradigm that makes WS systems resist change and it is not only the piped paradigm that makes it difficult to expand services by using small-scale, decentralized or off-grid technologies.

Apart from the power to decide on technology, the utilities have another type of power adding to the inertia of the traditional system. This power resides in the utilities obligation to oversee and control the abidance of health and environmental legislations imposed on both large and small service providers. These regulations put nearly the same demands on private or collective arrangements (70% of the purification level) as it does on municipal systems, and it is the municipality that checks if regulations are followed in decentralized and off-grid arrangements. The power to decide on rule abidance in health and environmental issues can be called 'institutional power.'

Thus, the municipality has two roles to play in the WS sector: both as system builder, with 'technological power,' and as overseer of legislation with 'institutional power.' Most often these two roles are managed by two separate municipal bodies which can create 'silo' effects within the organization and confusion on the local level of users and costumers.

The health dimension in water and sanitation was established from the very beginning. Already in the middle of the nineteenth century, the municipalities had the institutional power to oversee public health and when water systems, and later sewage, were established, control of health issues became even more central. Many municipalities built piped water and sewage systems, but it was not until 1955, in the *Public Water and Sewage Works Act* (SFS, 1955:314), that a municipal obligation was established for piped water and sewerage in urban areas. The law clarified the municipality's responsibility to provide water and sanitation, not just supervising public health.

Environmental issues due to pollution of rivers, lakes, and the sea were discussed early in the history of Swedish water and sanitation but not a lot of action was taken. However, the above-mentioned initiative to build sewage treatment plants in the 1960–1970s was a direct result of severe environmental degradation due to raw sewage being disposed straight into recipients. Accordingly, the environment was included in WS legislation at the end of the 1960s. In 1967, The Swedish Environmental Protection Agency was formed and in 1969 Sweden got its first national *Environmental Protection Act*. This law was the first in water and sanitation that dealt with <u>both</u> health and environmental issues in combination. One year later, in 1970, this development was codified in the new *Public Water and Sewage Works Act* (SFS, 1970:244) in a revision of the act of 1955.

In earlier research, colleagues and I have talked about the *critical interface* between the WS utility (the regime level) and unconnected user outside the grid (the local level) and that the regime tries to *bridge the interface* aiming to incorporate more and more of the unconnected users at the local level (Blomkvist and Nilsson, 2017; Blomkvist *et al.*, 2020; Karpouzoglou *et al.*, 2023; Blomkvist *et al.*, 2023).

This description is correct, but we have mainly focused on the pipes. We have been stuck in technology and (physical) system building aspects, concentrating on how the regime aims to connect the local level by traditional methods, that is expanding the well-tried piped network. The regime can use this expansion strategy because of its 'technological power.'

But there is another method to *bridge the interface*, based on 'institutional power.' By using its position as overseer and controller of legislation on health and environmental issues, the regime can expand its control over peripheral (private or collective) non-grid solutions. Although the law is stipulated by the state, the municipality has been delegated the 'institutional power' and the task to enforce the environmental legislation in water and sanitation. Thus, an institutional expansion strategy, through environmental legislation, is the second option for the regime to bridge the critical interface and to exert control over areas outside the grid.

What we see, apart from technical expansion, is the regime taking an ever-stronger grip over the WS sector by forcing non-grid property owners to abide to the letter of the law. The regime exerts technological and institutional power by acting in both its roles: as the technical system builder and the overseer of health and environmental legislation. In parallel to technical expansion through building of the grid and institutional expansion by environmental legislation, the building department of the municipality (Byggnadskontoret) also exerts, at least indirect, control over the WS sector by its municipal monopoly on building rights by stipulating which areas that are going to be incorporated in the future 'detailed building plan.' These areas are labeled service areas (verksamhetsområden) where the most common way to bridge the critical interface is by connecting the properties to the piped network.

Following from the above, I would argue for further research of the WS sector in Sweden focusing on how the regime have pushed the critical interface outwards from the center and expanded, what I would like to call, its 'influence domain.' I believe that the concept of an *influence domain*, i.e. the combined technical and institutional power of the municipalities, can be a useful distinction when discussing how to bridge the critical interface; to provide for water and sanitation in peripheral areas not yet connected to the grid and which obstacles need to be overcome in both a technical and an institutional sense. Furthermore, a discussion on the different power resources of the municipalities, technical and institutional, can be used in a future investigation on the various professional groups acting as carriers of technology and publicness in the WS sector. It is evident that the power of the municipal engineers has been challenged by professionals from environmental science. The history of a WS system, based on health considerations and civil engineering, gradually inoculated with environmental science, is yet to be written.

10.8 PUBLICNESS AND SYSTEMIC CHARACTERISTICS

In this section, I turn to some systemic characteristics that can help in understanding the different historical development of the three areas of municipal infrastructure under investigation. I argue that these varying systemic characteristics can explain differences in the articulation of publicness in roads and streets, water and sanitation.

Roads were from the very beginning connected to the values and ideological goals of movement, of both people and commodities. *Movement* was seen as something inherently beneficial for the country and its inhabitants and thus an obvious common good. Roads were also instruments for the highly valued concept of *connectivity*, which of course also applies to all modern transport and communication systems.

Using a concept from Jonsson (2000), roads and streets are a *communicative* inherently networked and gridded arrangements or systems with clear *positive network effects*. This means that a road gets its value by being connected to other roads in a network. Most often it would not be rational for an individual to build a road that only serves personal transport needs. Roads are not *local*, in a narrow sense. To be of real value, they need to connect point A to point B, to attach to other roads and cross a geographical space larger than the private domain. These characteristics have been apparent in Swedish road history for many hundred years, perhaps millennia.

Roads and streets, due to their inherent character as *communicative* networks, are a public good by definition, at least in most cases although there are modern exceptions such as motorways or toll roads which are *excludable* or when congestion turns the road into a *rivalrous* good. Nevertheless, historically roads and streets were clearly perceived as public goods. This perception is evident in the fact that roads and streets almost always have been financed by taxation. Attempts to introduce user fees, like in most other infrastructural systems, have met fierce resistance. The recent introduction of road pricing in Stockholm and Gothenburg was, and still is controversial, to say the least (Isaksson, 2008). It seems to me that the public character of Swedish roads and streets is firmly rooted.

From a systemic perspective, public and civic roads and streets have gradually become a well-aligned and cohesive infrastructural system with three integrated levels, exhibiting a strong *vertical integration*, with a distinct system builder managing each level. Sweden has a distinct state system builder who controls technical and institutional design on all three system levels. I would argue that, due to these systemic characteristics, the road and street network was articulated as a public concern long before it became an infrastructural *system* in our modern meaning and that the authorities showed a clear public ambition long before they really became system builders.

The area of drinking water is a bit peculiar because water provision was in a sense a public matter and water by its nature a public good according to old customs, but there were no strong initiatives to articulate drinking water as a public responsibility. But this changed dramatically when water provision was connected to the value of 'sanitary economy' based on a new perception of 'prophylactic health.' Adding to this, the new industrialists also realized that factory workers were a valuable production factor that needed to be kept in good health to be able to create surplus value.

Modern drinking water provision is a *distributive* system (Jonsson, 2000) with a strong *local* character, due to the locality of water resources and thus exhibiting weak positive network effects. Drinking water was for a long time regarded as an exclusively private and local matter, and in the countryside and in sparsely populated areas, this did not really change until the first half of the twentieth century. In towns, population density spurred more cooperation in water management such as common wells and water selling, thus increasing network effects. However, the local character of the resource prevailed. Although a gridded system locally, water provision never became a nationally, well-aligned and cohesive infrastructural system as roads and streets. Water provision does not exhibit a strong *vertical integration*. In the present, water provision, and to some extent sewage, has lost some of its local qualities as water (and sewage) is transported quite long distances between cooperating municipal WS organizations. This means that water provision is becoming a *horizontally integrated* system.

Provision of drinking water has historically and for a long period mostly been defined as a *private* good concerning the possibility to exclude other users. It is quite easy to put up a fence around your private well. Rivalry over drinking water has not been an important issue in Sweden. If there is no apparent shortage, water is not a rivalrous good. Perhaps the relative abundance of water in Sweden is a reason why publicness in drinking water has not really been strongly articulated.

Inner and outer sanitation are *accumulative* arrangements or systems (Jonsson, 2000) with less visible positive network effects, at least in sparsely populated areas. It is not obvious for the individual that cooperation with others is beneficial. Waste and excrement can be managed *locally* within your own property, and this is exactly what has been the historical norm. This was true also for Swedish towns when they were relatively small. However, with growing population density, negative externalities such as stench become apparent. The need for cooperation increased, first in outer and later in inner sanitation, became pressing and positive network effects became gradually more apparent.

Thus, sanitation was eventually articulated as a public responsibility, but it was only inner sanitation, that is wastewater and latrine management, that turned into a gridded system. The local character prevailed though, just like in water provision and although a gridded system locally, sewage never became a nationally, well-aligned and cohesive infrastructural system. Sewage, again just as water, did not exhibit a strong *vertical integration*, but has become more and more *horizontally integrated* because of cooperating municipal organizations.

Excrement removal and wastewater disposal have historically and for a long period mostly been defined as a *private* good in the meaning that the result, cleanliness, was seen as something only benefiting the private individual and the household. With increasing population density, when more people were affected, and a new perception of health and sickness, cleanliness, the absence of filth and stench became articulated as a public responsibility which finally resulted in publicly managed outer sanitation and underground sewerage. Thus publicness in sanitation was articulated a bit stronger than in water provision, but not as strong as in roads and streets.

As a concluding remark, it must be noted that even if water and sewage gradually turned into a unified system in the Swedish towns and eventually all over Sweden, some institutional differences persisted between the two. The historical legacy of arrangements for water provision and sewage removal is evident in which municipal organizations were put in charge of the systems. In Stockholm, for example, drinking water was managed by specialized municipal bodies, that is the board of the water works in cooperation with the financial board (Drätselnämnden) up until the 1920s when water management was merged with another infrastructure, the gas works. Sewage, on the other hand, was first managed by the financial board and its divisions dealing with buildings and street management. The connection between sewerage and street management had its origin in the heritage from street cleaning and stormwater management, which belonged to municipal bodies dealing with street keeping and waste removal. This legacy survived until 1974 when water and sewage were joined in a common organization, the Water and Sewage Works (today the municipal company Stockholm Water and Waste). The same organizational division seems to have been the norm all over Sweden, and even today some of this legacy is still alive. Piped sewage is normally managed by the municipal water and sewage utility while the emptying of latrine from three-chamber wells outside the grid is handled by the street or solid waste divisions. Thus, the same human refuse is managed as sewage by one municipal body and as solid waste by another.

Building an infrastructure system requires large initial investments that few, if any individuals or private companies can afford. The construction has a long lead time, and it is not an easy task to build a system gradually because it is hard to get return on investments until the whole system is completed and all users are connected. Because of these reasons, the state or the municipality has often stepped in as a guarantor of construction and maintenance, and most Swedish infrastructure systems have been nationalized, although many were originally built by users and appropriators in local cooperatives (electricity, telephone, etc.). Since the end of the twentieth century, we have witnessed a re-regulation of the infrastructure systems which resulted in private companies taking over ownership, operation, and maintenance.

In the history of roads and streets, the state and the cities have since medieval times shown a public engagement, but they cannot really be called system builders. In the beginning of the twentieth century though, the state and the municipalities had enough power and resources to act as system builder and create a true infrasystem. In water and sanitation, we do not see this clear central system builder role. As has been shown, the state and municipal authorities had the ambition and they surely tried to manage sanitation, but most often their efforts were fruitless. In water provision, public involvement has been less apparent and came later. We still do not see a state level system builder in water and sanitation.

With the risk of repeating myself, I want to underline that service arrangements such as road and street keeping, water provision, and sanitation can well be a public concern (undertaking and responsibility) and yet not a real infrastructural system. These three areas of public service grew slowly, and in various paces, into fully developed infrasystems. They were gradually *systemized*, and their systemic characteristic became more apparent over time. In some, like roads, and to some extent sanitation, publicness came before *systemization*. In others, like piped water and sewage, full publicness developed in conjunction with *systemization*.

10.9 PROUD SYSTEM AND COMMUNITY BUILDERS

As discussed above, when roads and streets and water and sanitation gradually evolved into infrastructural systems, when the former service arrangements were systemized, modern technology and management was introduced by a new set of actors taking on the role of system builders or system promoters alongside the municipalities and the state. These new actors changed the articulation of publicness in infrastructure connecting it to the creation of a Swedish well fare state based on industrial growth. From now on, publicness was associated with future visions of modern technology and infrastructural services for all. I have called these actors *carriers of technology and publicness*. Moreover, these actors helped in creating a *systems culture* based on engineering science in both sectors.

In this last section, I turn to an aspect often forgotten in the history of infrastructure which I believe was important in the systemization of both roads and streets and water and sanitation: the professionalization process of civil engineers and their self-image of being proud system and community builders. Based on earlier research (Blomkvist, 2001), I want to highlight the societal role and significance of Swedish engineers in the so-called second industrialization in Sweden during the first two decades of the twentieth century, a period characterized by a general technological and scientific optimism that in many ways resembles the decades after WWII. The engineering community wanted to establish their profession as independent scientific discipline. The engineer would no longer be just a technician who practically applied the achievements of the natural scientist but a scientist in *his* own right (almost all engineers were men at the time). This general *scientification* of engineering can be exemplified by the establishment of the Academy of Engineering Sciences in 1919 and the introduction of the technical doctorate level at the Royal Institute of Technology (KTH) in 1927. The road and water engineers' professionalization efforts were therefore in line with other engineering groups at this time.

The professionalization process was also intimately connected to the strong team spirit that existed (and still exists) among the road and water builders. The most important components of the 'Roads and Water builder spirit' were the already mentioned military origins and the common educational background at the technical universities which formed the basis for the profession's strong 'masculine coding' and the engineer's identity as a 'community builder.' The road and water builders were part of a 'brotherhood,' a male camaraderie, with a clear identity originating at the military institutions where the first civil engineers were educated. When engineering education was transferred to technical colleges, these military and masculine coded norms remained important. In a historical essay on the student union at KTH, the brotherhood is described like this: 'Within every profession there is something of Freemasonry, for better or for worse. We call it community spirit and feel proud to belong to the community that is the bearer of this subtle feeling. It is born out of traditions, which live on at the university' (Widegren, 1967 in Blomkvist, 2001).

Alongside this 'freemasonry,' there was a strong belief among the engineers that they had a special role in the creation of the Swedish nation based on their experience of working directly in the service of society. Their self-image included the role of the pioneer and 'trailblazer' that cleared the ground for the industrial revolution and with a healthy 'navvy spirit' built the railways and constructed the material foundation for the well fare state. It was the road and water builder who founded society in a concrete sense and the trailblazer from the latter half of the nineteenth century was gradually transformed into the community builder from the second half of the twentieth century. They saw themselves as tradition-bearers in one of the world's oldest professions and maintained respect for the strength and skill of the hand long after machines had gained entry.

Thus, important components of the road and water builder's systems culture have been a technical-scientific approach and a common value base created by a similar educational background resulting in a self-image of being a proud system and community builder. Sometimes this last aspect is forgotten when discussing technical and scientific experts and their influence over large infrastructural or industrial projects. The system culture is relevant because it contributes to system inertia and a reluctance to change. Researchers, not least me, often critically focus on the engineers and other experts' power in defining critical problems and propose solutions to these. If these actors are influenced by mindsets and beliefs based on the system culture, there is a great risk that they will only see one kind of solution to the problems at hand. Radical proposals such as completely changing the direction of the system do not get sufficient attention.

Nevertheless, the image of the proud system and community builder has an upside that must not be forgotten. This positive self-image was crucial for the system builders of infrastructure in Sweden from the end of the nineteenth century and onwards. They really felt that they contributed to a better future for the nation and for its inhabitants. They laid the ground for a welfare state where universal access to public streets, roads, water and sanitation was the final goal. Roads and water builders, in their own perspective, were truly building the future society and accepted all the challenges offered by this endeavor. It is my belief that the challenges facing municipal (and national) infrastructure today, mainly because of climate effects, would need a similar self-image among influential actors. The future management of global warming and climate change mitigation in our infrastructural systems could certainly benefit from a new generation of proud system and community builders.

doi: 10.2166/9781789063981_0123

References

The reference list includes only literature. Primary sources, contemporary print, and web sites are left out. The list is condensed and adapted to an international audience. Nevertheless, I have added more items than cited directly in the text. The reason is that I want to give the reader an overview of the large literature in the field. Please note that the translations of Swedish titles are done by me as a service to foreign readers. Therefore, they may not exactly match bibliographic references.

A complete list of *all* references can be found in the research report which is the foundation for this book: *Research report and excerpts on the history of municipal streets, water and sanitation in Sweden* (Blomkvist, 2023a) on this website: Publikationer – Mistra InfraMaint. The report contains lots of raw data in the form of excerpts and quotes from literature and primary sources and all relevant information in the form of footnotes. It has a similar structure as the book, so it should be quite easy to find the appropriate reference, side numbers, etc.

Alm J. and Paulsson A. (2023). Uncommon commons. In: *Public Participation in Transport in Times of Change*, L. Hansson, C. Hedegaard Sørensen and T. Rye (eds), 1st edn, Emerald Group Publishing Limited, Bingley, England, pp. 97–114.

Alm J., Paulsson A. and Jonsson R. (2021). Capacity in municipalities: infrastructures, maintenance debts and ways of overcoming a run-to-failure mentality. *Local Economy*, **36**(2), 81–97.

Anderberg S. (1986). *Stockholms vattenförsörjning genom tiderna (History of Stockholm's water supply)*. Stockholm.

Andersson A. (1971). *Svenska vattenledningar och vattenreservoarer 1869–1910 (Swedish waterpipes and reservoirs 1869–1910)*, C-uppsats i konstvetenskap. Uppsala universitet, Uppsala.

Andersson J. (2013). *Karlskrona vattenverk 150 år: stadens och kommunens VA-historia 1863–2013 (Karlskrona waterworks 150 years: the WS history of the city 1863–2013)*. Karlskrona.

Andersson-Skog L. (1993). *Såsom allmänna inrättningar till gagnet, men affärsföretag till namnet. SJ, järnvägspolitiken och den ekonomiska omvandlingen efter 1920 (As public establishments for the benefit of all, but business enterprises in name. SJ, railway politics and the economic transformation post 1920)*. Surte.

Andersson-Skog L. (1998). De osynliga användarna: telefonen och vardagslivet 1880–1995 (The invisible users: telephones and everyday life). In: *Den konstruerade världen: Tekniska system i historiskt perspektiv (The constructed world: technical systems in a historical perspective)*, P. Blomkvist and A. Kaiser (eds), Brutus Östlings Bokförlag Symposium, Stockholm, pp. 277–298.

Andersson-Skog L. and Ottosson J. (2018). *Stat och marknad i historiskt perspektiv: från 1850 till i dag (State and market in a historical perspective: from 1850 until today)*, Första upplagan. Stockholm (Dialogos).

Aronsson P. (1993). *Bönder gör politik: Det lokala självstyret som social arena i tre smålandssocknar, 1680–1850 (Farmers doing politics: the local rule as social arena in three smaller municipalities, 1680–1850)*. Lund University Press.

Bäckman H. (1984). *Avloppsledningar i svenska tätorter i ett historiskt perspektiv (Sewage in Swedish urban areas in a historical perspective)*, Meddelande nr 74, Chalmers tekniska högskola, Göteborg.

Benes C. (2009). Whose SPQR? Sovereignty and semiotics in medieval Rome. *Speculum*, **84**(4), 874–904, https://doi.org/10.1017/S0038713400208130

Bennich A., Engwall M. and Nilsson D. (2023). Operating in the shadowland: why water utilities fail to manage decaying infrastructure. *Utilities Policy*, **82**, 101557, https://doi.org/10.1016/j.jup.2023.101557

Binnie G. M. (1981). *Early Victorian Water Engineers*. Telford.

Björkman B. (1967). *Väg- och vattenbyggaren i ett föränderligt samhälle (The road and water builder in a changing society)*. Jubileumsskrift 125 år Väg- och vattenbyggaren, Stockholm.

Björkman J. (2020). *"Må de herrskande klasserna darra": Radikal retorik och reaktion i Stockholms press, 1848–1851 ("May the ruling classes shiver": radical rethoric and reaction in Stockholm Press, 1848–1851)*. Doctoral dissertation, Gidlunds förlag, Hedemora.

Bjur H. (1988). *Vattenbyggnadskonst i Göteborg under 200 (The of water building in Gothenburg during 200 years)*. Göteborgs VA-verk, Göteborg.

Bjur H. and Malbert B. (1988). *Under staden: perspektiv på kommunal infrastruktur (Under the city: perspectives of municipal infrastructure)*. Statens råd för byggnadsforskning, Stockholm.

Black M. and Fawcett B. (2008). *The Last Taboo: Opening the Door on the Global Sanitation Crisis*. Earthscan, London.

Blomkvist P. (2001). *Den goda vägens vänner: väg- och billobbyn och framväxten av det svenska bilsamhället 1914–1959 (Friends of good roads: the road and car lobby, and the growth of the Swedish car society)*. Diss., Univ. Brutus Östlings Bokförlag Symposium, Stockholm.

Blomkvist P. (2004). Transfering technology – shaping ideology. American traffic engineering, experts and commercial interests in the establishment of a Swedish, and European, car society in the post war period. *Comparative Technology Transfer and Society*, **2**(3), 273–302.

Blomkvist P. (2010). *Om förvaltning av gemensamma resurser: Enskild väghållning och allmänningens dilemma i svensk historia 1200–2010 (Managing common pool resources: road keeping and the dilemma of the commons in Swedish history*

1200–2010), TRITA-IEO 2010:06, Division of Industrial Dynamics, KTH (Royal institute of Technology), Stockholm.

Blomkvist P. and Emanuel M. (2020). Regulating innovation: Utility vs. leisure in Swedish moped history, 1952–1961. *Technology and culture*, **61**(3), 815–842. Johns Hopkins University Press.

Blomkvist P. (2023a). *Research Report and Excerpts on the History of Municipal Streets, Water and Sanitation in Sweden*, RISE.

Blomkvist P. and Johansson P. (2016). Systems thinking in industrial dynamics. In: *A Dynamic Mind. Perspectives on Industrial Dynamics in Honour of Staffan Laestadius*, P. Blomkvist and P. Johansson (eds), Division of Sustainability and Industrial Dynamics, Department of Industrial Economics and Management, KTH (Royal Institute of Technology), Stockholm, pp. 45–75.

Blomkvist P. and Kaiser A. (1998). Den konstruerade världen: Tekniska system i historiskt perspektiv (The constructed world: technical systems in a historical perspective). Brutus Östlings Bokförlag Symposium, Stockholm.

Blomkvist P. and Larsson J. (2013). An analytical framework for common-pool resource-large technical system (CPR-LTS) constellations. *International Journal of the Commons*, **7**(1), 113–139, https://doi.org/10.18352/ijc.353

Blomkvist P. and Nilsson D. (2017). On the need for system alignment in large water infrastructure: understanding infrastructure dynamics in Nairobi, Kenya. *Water Alternatives*, **10**(2), 283–297.

Blomkvist P., Nilsson D., Joma B. and Sitoki L. (2019). Bridging the critical interface: ambidextrous innovation for water provision in Nairobi's informal settlements. *Technology in Society*, **60**(2020), 1–12.

Blomkvist P., Karpouzoglou T., Nilsson D. and Wallin J. (2023). Entrepreneurship and alignment work in the Swedish water and sanitation sector. *Technology in Society*, **74**, 102280, https://doi.org/10.1016/j.techsoc.2023.102280

Boccaletti G. (2022). *Water – A Biography*. Random House USA, New York, USA.

Bokholm R. (1995). Städernas handlingsfrihet: en studie av expansionsskedet 1900–1930 (City freedom of action: a study of the state of expansion 1900–1930). Diss., Universitet Lund, Lund.

Braadbaart O. (2009). North–South transfer of the paradigm of piped water: the role of the public sector in water and sanitation services. In: *Water and Sanitation Services Public Policy and Management*, J. E. Castro and L. Heller (eds), Routledge, London, pp. 99–113.

Brown R. R. and Farrelly M. A. (2009). Delivering sustainable urban water management: a review of the hurdles we face. *Water Science and Technology*, **59**(5), 839–846, https://doi.org/10.2166/wst.2009.028

Bruun C. (2013). Water supply, drainage, and watermills. In: *The Cambridge Companion to Ancient Rome (Cambridge Companions to the Ancient World)*, P. Erdkamp (ed.), Cambridge University Press, Cambridge, England. pp. 297–314.

Cettner A., Söderholm K. and Viklander M. (2012). An adaptive stormwater culture? Historical perspectives on the status of stormwater within the Swedish urban water system. *Journal of Urban Technology*, **19**(3), 25–40, https://doi.org/10.1080/10630732.2012.673058

Christensen J. (2003). Enskilda avlopp – miljöbalken har ändrat de rättsliga förutsättningarna (Small scale sewage – the environmental code has changed the regulatory conditions). In *Miljörätten i förändring – En antalogi (Environmental law in transition – an anthology)*, G. Michanek and U. Björkman, **36**. Iustus, Uppsala, Sverige, 1–42.

Christensen J. (2015). *Juridiken kring vatten och avlopp. En översiktlig genomgång av juridiken kring dricksvattenförsörjning samt avledning och rening av spillvatten* och dagvatten (Legal matters in water and sewage. A review of the legal factors surrounding the supply of drinking water and diversion and cleaning of wastewater), Havs- och vattenmyndighetens rapport 2015:15, Stockholm.

Ciriacono S. (1998). Land Drainage and Irrigation. Ashgate Publishing Ltd., Aldershot.

Correljé A. and Schuetze T. (2012). Decentral water supply and sanitation. In: Inverse Infrastructures: Disrupting Networks from Below, T. M. Egyedi and D. C. Mehos (eds), Edward Elgar, Cheltenham, pp. 161–186.

Cronström A. (1986). Stockholms tekniska historia, 3: Vattenförsörjning och avlopp (The technical history of Stockholm, 3: water supply and sewage). Liber Förlag, Stockholm.

Crow J. (2012). Ruling the waters: managing the water supply of Constantinople, AD 330–1204. Water History, 4, 35–55, https://doi.org/10.1007/s12685-012-0054-y

Dalgaard C. J., Kaarsen N., Olsson O. and Selaya P. (2022). Roman roads to prosperity: persistence and non-persistence of public infrastructure. Journal of Comparative Economics, 50(4), 896–916, https://doi.org/10.1016/j.jce.2022.05.003

Dellapenna J. W. and Gupta J. (eds). (2009). The Evolution of the Law and Politics of Water. Springer Netherlands, Dordrecht.

Deming D. (2020). The aqueducts and water supply of ancient Rome. Groundwater, 58(1), 152–161, https://doi.org/10.1111/gwat.12958

Dixon N. (2007). Healthy water from an indigenous mauri perspective. Environmental History of Water, 475. IWA Publishing, London, England.

Drakman A. (2018). När kroppen slöt sig och blev fast: varför åderlåtning, miasmateori och klimatmedicin övergavs vid 1800-talets mitt (When the body closed and solidified: why bloodletting, miasma theory, and climate medicine was abandoned in the mid 1800). Diss., Uppsala universitet, Uppsala.

Drangert J.-O. (1991). Svensk vattenhistoria (Swedish history of water). Universitetet, Tema Vatten, Linköping.

Drangert J.-O. (1993). Who Cares About Water? A Study of Household Water Development in Sukumaland, Tanzania. Linköping University, Linköping.

Drangert J.-O. and Hallström J. (2003). Den urbana renhållningen i Stockholm och Norrköping: från svin till avfallskvarn? (Urban sanitation in Stockholm and Norrköping: From pigs to a waste mills). Bebyggelsehistorisk tidskrift, 44, 7–24.

Drangert J.-O. and Löwgren M. (2005). Förändring eller kontinuitet?: faktorer som påverkat va-systemens utveckling i Linköping och Norrköping under perioden 1960–1990 (Change or continuity?: Factors that have changed the development of WS-systems in Linköping and Norrköping between 1960–1990). Chalmers tekniska högskola, Göteborg.

Drangert J.-O., Nelson M. C. and Nilsson H. (2002). Why did they become pipe-bound cities? Early water and sewerage alternatives in Swedish cities. Public Works Management & Policy, 6(3), 172–185, https://doi.org/10.1177/1087724X0263003

Dufwa A. (1985). Trafik, broar, tunnelbanor, gator (Traffic, bridges, subways, streets), Stockholms tekniska historia 1. Liber Förlag, Stockholm.

Dufwa A. and Pehrson M. (1989). Avfallshantering och återvinning (Waste management and recycling). Stockholm.

Edquist C. and Edqvist O. (1979). Social carriers of techniques for development. Journal of Peace Research, 16(4), 313–331, https://doi.org/10.1177/002234337901600403

Edwards P. N. (2003). Infrastructure and modernity: force, time, and social organization in the history of sociotechnical systems. In: Modernity and Technology, T. Misa, P. Brey and A. Freeberg (eds), MIT Press, Cambridge, MA, pp. 185–225.

Egyedi T. M. (2012). Disruptive inverse infrastructures: conclusions and policy recommendations. In: Inverse Infrastructures: Disrupting Networks from Below, T. M. Egyedi and D. C. Mehos (eds), Edward Elgar, Cheltenham, pp. 241–266.

Ehn W. (1982). *Byordningar från Mälarlänen (Village by-laws from the Mälar counties)*. Stockholms, Södermanlands, Uppsala and Västmanlands county, collected and distributed by Wolter Ehn. Writings provided through the dialect and folk memory archives in Uppsala Ser. B:16, Uppsala.

Emanuel M. (2012). Trafikslag på undantag: cykeltrafiken i Stockholm 1930–1980 (Traffic type put on exception: cycling traffic in Stockholm 1930–1980). Diss., Kungl. tekniska högskolan, Stockholm.

Fischer C. S. (1988). Gender and the residential telephone, 1890–1940: technologies of sociability. *Sociological Forum*, 3(2), 211–233, https://doi.org/10.1007/BF01115291

Fishman C. (2011). *The Big Thirst: The Secret Life and Turbulent Future of Water*. Free Press, New York.

Forsberg A. (2013). Local responses to structural changes: collective action for rural communities in Sweden. In: *Social Capital and Rural Development in the Knowledge Society*, H. Westlund and K. Kobayashi (eds), Edward Elgar, Cheltenham, pp. 247–272.

Forsell H. (2003). Hus och hyra: fastighetsägande och stadstillväxt i Berlin och Stockholm 1860–1920 (House and rent: Property ownership and urban growth in Berlin and Stockholm 1860–1920). Diss., Stockholm Universitet, Stockholm.

García, P. A. (2007) Water and environment in one indigenous region of Mexico (chapter 27: pp. 411–428) in Juuti, P, Katko, T, Vuorinen, H (eds.) (2007). *Environmental history of water: global views on community water supply and sanitation*. London: IWA Publishing

Geels F. W. (2006). The hygienic transition from cesspools to sewer systems (1840–1930): the dynamics of regime transformation. *Research Policy*, 35(7), 1069–1082, https://doi.org/10.1016/j.respol.2006.06.001

Goubert J.-P. (1989). *The Conquest of Water: The Advent of Health in the Industrial Age*. Polity, Cambridge, UK.

Grönvall A. (2018). *Vägar till hållbara vattentjänster [SOU 2018:34] (Roads to sustainable water services)*. Elanders Sverige AB, Stockholm.

Gullberg A. (1998). Nätmakt och maktnät: Den nya kommunaltekniken (The power of the grid and power grids: the new municipal technology). In: *Stockholm 1850–1920. Den konstruerade världen: Tekniska system i historiskt perspektiv* (The constructed world: Technological systems in a historical perspective), P. Blomkvist and A. Kaiser (eds), Brutus Östlings Bokförlag Symposium. Höör, Sverige, pp. 105–129.

Gullberg A. (2001). *City: drömmen om ett nytt hjärta: moderniseringen av det centrala Stockholm 1951–1979 (City: the dream of a new heart: modernization of central Stockholm 1951–1979)*. Stockholmia, Stockholm.

Gunn S., Butler R., De Block G., Høghøj M. and Thelle M. (2022). Cities, infrastructure and the making of modern citizenship: the view from north-west Europe since c. 1870. *Urban History*, pp. 1–19.

Hallenberg M. (2018). *Kampen om det allmänna bästa: konflikter om privat och offentlig drift i Stockholms stad under 400 år (The fight for the common good: conflicts between private and public management in Stockholm for 400 years)*. Nordic Academic Press, Lund.

Hallenberg M. and Linnarsson M. (2016). Vem tar bäst hand om det allmänna? Politiska konflikter om privata och offentliga utförare 1720–1860 (Who to manage the public good? Political conflicts surrounding private and public execution). *Historisk tidskrift*, 136(1), 32–63.

Hallenberg M. and Linnarsson M. (2017). The quest for publicness political conflict about the organisation of tramways and telecommunication in Sweden, c. 1900–1920. *Scandinavian Economic History Review*, 65(1), 70–87, https://doi.org/10.1080/03585522.2016.1258007

Halliday S. (2001[1999]). *The Great Stink of London: Sir Joseph Bazalgette and the Cleansing of the Victorian Metropolis*. Sutton, Stroud.

Hallström J. (2003). *Constructing a Pipe-Bound City: A History of Water Supply, Sewerage, and Excreta Removal in Norrköping and Linköping, Sweden, 1860–1910*. PhD thesis, University of Linköping, Linköping.

Hallström J. (2005). Technology, social space, and environmental justice in Swedish cities: water distribution to suburban Norrköping and Linköping, 1860–90. *Urban History*, **32**(3), 413–433, https://doi.org/10.1017/S0963926805003214

Hallström J. and Melosi M. V. (2022). History of technological change in urban wastewater management, 1830–2010. *Routledge Handbook of Urban Water Governance*, pp. 163–172. Routledge, Abingdon, England.

Hamlin C. (1992). Edwin Chadwick and the engineers, 1842–1854: systems and anti-systems in the pipe-and-brick sewers war. *Technology & Culture*, **33**(4), 680–709.

Hamlin C. (1998). *Public Health and Social Justice in the Age of Chadwick. Britain, 1800–1854*. Cambridge, UK.

Hamlin C. (2009). "Cholera forcing" the myth of the good epidemic and the coming of good water. *American Journal of Public Health*, **99**(11), 1946–1954, https://doi.org/10.2105/AJPH.2009.165688

Hardwick L. (2003). *Reception Studies. Greece and Rome: New Surveys in the Classics (33)*. Oxford University Press, Oxford, UK.

Heddelin B. (red.). (1991). *VÄGAR. Dåtid, Nutid, Framtid (ROADS. Past, Present, Future)*. Stockholm.

Hehmeyer I. (2007). Water, lifeline of the city of Ghayl ba Wazir, Yemen. In: *Environmental History of Water: Global Views on Community Water Supply and Sanitation*, P. Juuti, T. Katko and H. Vuorinen (eds), IWA Publishing, London, pp. 197–212.

Heino O. and Anttiroiko A.-V. (2014). *Enabling and Integrative Infrastructure Policy: The Role of Inverse Infrastructures in Local Infrastructure Provision with Special Reference to Finnish Water Cooperatives*, MPRA Paper No. 60276, University of Tampere.

Hodge, A. Trevor: "Part 1, "Water supply"; chapter 3 "Wells" (p 29–34) and chapter 5 "Aqueducts" (p. 39–66) in Wikander, Örjan (red.) (2000). *Handbook of ancient water technology*. Leiden: Brill.

Höjer T. (1953). Stockholms stads drätselkommission 1814–1864 och Börs-, bro- och hamnbyggnadskommitterade 1815–1846 (Stockholm' financial board 1814–1864 and the market-, bridge-, and port building committee 1815–1846). Diss., Stockholms högskola, Stockholm.

Höjer T. (1967). *Sockenstämmor och kommunalförvaltning i Stockholm fram till 1864 (Parrish Councils and municipal management in Stockholm until 1864)*. Stadsarkivet, Stockholm.

Holmquist H. (1961). *Vattenverket 100 år, minnesskrift (Waterworks 100 years, commemorative script)*. Stockholm Waterworks in Collaboration with Stockholms Stadsmuseum, Stockholm.

Hughes T. P. (1987). The evolution of large technological systems. In: *The Social Construction of Technological Systems. New Directions in the Sociology and History of Technology*, W. E. Bijker, T. P. Hughes and T. Och Pinch (eds), The MIT Press, Cambridge, MA, pp. 51–82.

Hughes T. P. (1988). *Networks of Power: Electrification in Western Society (1880–1930)*. The John Hopskins University Press, London, UK.

Hukka J. and Katko T. (2009). Complementary paradigms of water and sanitation services: lessons from the Finnish experience. In: *Water and Sanitation Services:*

Public Policy and Management, E. Castro and L. Heller (eds), Earthscan, Oxford, England, pp. 153–172.

Isaksson K. (red.). (2008). *Stockholmsförsöket: en osannolik historia (The Stockholm trial: an unbelievable story)*. Stockholmia, Stockholm.

Isgård E. (1998). *I Vattumannens tecken: svensk VA-teknik från trärör till kväverening (In the sign of Aquarius: Swedish WS technology from wood pipes to nitrogen purification)*. Ohlson & Winnfors, Örebro.

Isgård E., Mjöberg H. and Rundgren L. (1987). *VBB-perspektiv 1987: teknik med perspektiv (VBB-perspective 1987: Technology in a perspective)*. Vattenbyggnadsbyrån (VBB), Stockholm, Sverige.

Jakobsson E. (1996). *Industrialisering av älvar. Studier kring svensk vattenkraftutbyggnad 1900–1918 (Industrialisation of rivers. Studies on Swedish water power construction 1900–1918)*. Diss., Historiska institutionen i Göteborg, nr 13, Göteborg.

Jakobsson E. (1999). Introduktion av WC i Stockholm: ett vattensystemperspektiv på staden (Introduction of WC in Stockholm: a water systems perspective of the city), Polhem. *Tidskrift för teknikhistoria*, **17**(2–4), 118–139.

Johansson I. (1991). *Stor-Stockholms bebyggelsehistoria. Markpolitik, planering och byggande under sju sekler (The history of settlement in greater Stockholm. Land politics, planning and construction during seven centuries.)*. Värnamo.

Johansson B. (1997). *Stadens tekniska system. Naturresurser i kretslopp (The technical systems of the city. Natural resources in circulation)*. Byggforskningsrådet, Stockholm.

Jonsson D. (2000). Sustainable infrasystem synergies: a conceptual framework. *Journal of Urban Technology*, **7**(3), 81–104, https://doi.org/10.1080/713684136

Juuti P. and Katko T. (2004). *From a Few to All – Long Term Development of Water and Sanitation Services in Finland*. Tampere.

Juuti P. and Katko T. (2005). *Water, Time, and European Cities: History Matters for the Futures*. WaterTime, Tampere.

Juuti P., Katko T., Persson K. and Rajala R. (2009). Shared history of water supply and sanitation in Finland and Sweden, 1860–2000. *Vatten*. **3**, 165–175.

Juuti P. K. and Vuorinen T. H. (2007). *Environmental History of Water: Global Views on Community Water Supply and Sanitation*. IWA Publishing, London, UK.

Juuti P., Katko T. and Hukka J. (2007) Privatisation of water services in historical context, mid-1800s to 2004 in Juuti, P., Katko, T. and Vuorinen, H. (eds.) (2007). *Environmental history of water: global views on community water supply and sanitation*. London: IWA Publishing. pp. 235–257.

Kaijser A. (1986). Stadens ljus: etableringen av de första svenska gasverken (City lights: the establishment of the first Swedish gas works). Diss., Universitet Linköping, Linköping.

Kaijser A. (1994). *I fädrens spår: den svenska infrastrukturens historiska utveckling och framtida utmaningar (Following in the footsteps of father: historical development and future challenges of Swedish infrastructure)*. Carlsson, Stockholm.

Kaijser A. (1999). Den hjälpsamma handen: Om den instutionella utformningen av svenska infrasystem (The helping hand: the institutional design of Swedish infrasystems). *Historisk tidskrift, Tema: Teknikhistoria*, **1999**(119), 397–434.

Kaiserfeld T. and Kaijser A. (2021). Changing the system culture: mobilizing the social sciences in the Swedish nuclear waste system. *Nuclear Technology*, **207**(9), 1456–1468, https://doi.org/10.1080/00295450.2020.1832815

Karlsson A. (2021). *Vatten: En historia om människor och civilisationer (Water: a history of people and civilisations)*. Svenska Historiska Media Förlag, Lund, Sverige.

Karpouzoglou T., Vij S., Blomkvist P., Juma B., Narain V., Nilsson D. and Sitoki L. (2023). Analysing water provision in the critical interface of formal and informal urban water regimes. *Water International*, **48**(2), 202–216,Routledge.

Katko T. (1992). *Evolution of Consumer-Managed Water Cooperatives in Finland, with Implications for Developing Countries.* Water International. Routledge, Abingdon, England.

Katko T. (2000). Long-term development of water and sewage services in Finland. *Public Works Management & Policy*, **4**(4), 305–318, https://doi.org/10.1177/1087724X0044005

Katko T., Juuti P. and Pietila P. (2006). Key long-term strategic decisions in water and sanitation services management in Finland, 1860–2003. *Boreal Environment Research*, **11**, 389–400.

Katko T., Juuti P. and Rajala R. (2009). Writing the history of water services. *Physics and Chemistry of The Earth*. **34**, 156–163.

Kilander S. (1991). Den nya staten och den gamla. En studie i ideologisk förvandling (The new state and the old. A study in ideological change). *Acta Universitatis Upsaliensis, Studia Historica Upsaliensia*, **164**.

Kjellén M. (2006). *From public pipes to private hands: water access and distribution in Dar es Salaam, Tanzania.* PhD dissertation, Acta Universitatis Stockholmiensis. Stockholm, Sverige.

Landin B. and Henrikson L. (2022). Vatten land: om våtmarkens roll i det utdikade landskapet. (Water and land: On the role of wetlands in the diked-out landscape). Max Ström, Stockholm.

Latour B. (1987). *Science in Action: How to Follow Scientists and Engineers Through Society.* Harvard University Press, Cambridge, MA.

Lawhon M., *et al.* (2017). *Thinking Through Heterogeneous Infrastructure Configurations, Urban Studies*, **55**(4), 720–732.

Lay M. G. (1999). *Ways of the World: A History of the World's Roads and of the Vehicles That Used Them.* Rutgers University Press. New Brunswick, USA.

Lindberg E. (2015). The Swedish lighthouse system 1650–1890: private versus public provision of public goods. *European Review of Economic History*, **19**(4), 454–468, https://doi.org/10.1093/ereh/hev015

Lindberg E. (2022). *Välfärdens vägar: organiseringen av vägunderhållet i Sverige 1850–1944 (The roads of welfare: organisation of road maintenance in Sweden 1850–1944).* Nordic Academic Press, Lund.

Linnarsson M. and Hallenberg M. (2016). Urban space, private business and the common good. The politics of the street in early 20th century Stockholm. Abstract from conference presentation: reinterpreting cities: European Association for Urban History Conference, August 24–27, Helsinki, Finland.

Linnarsson M. and Hallenberg M. (2020). The shifting politics of public services: discourses, arguments, and institutional change in Sweden, c. 1620–2000. *Journal of Policy History*, **32**(4), 463–486, https://doi.org/10.1017/S0898030620000184

Löfsten H. (1992). Underhåll av kommunal infrastruktur: principer för planering, prissättning och finansiering (Maintenance of municipal infrastructure: *principles* for planning, pricing and financing). Diss., Universitet Göteborg, Göteborg.

Lundgren L. (1974). Vattenförorening. Debatten i Sverige 1890–1921 (Water pollution. The Swedish debate in Sweden 1890–1921). *Bibliotheca historica Lundensis*, **30**, (PhD Dissertation) Lunds universitet, Lund, Sverige.

Lundin P. (2008). *Bilsamhället: ideologi, expertis och regelskapande i efterkrigstidens Sverige (The car society: ideology, expertise and rule setting in post war Sweden).* Diss., Kungliga tekniska högskolan, Stockholm.

Malm A. and Löfdahl G.-M. (2020). Engaging stakeholders for improved IAM implementation. *Water Practice and Technology*, **15**(2), 350–355.

Mårald E. (2002). Vårt bästa guld: Agrara perspektiv på urban teknik från 1800-talets mitt till dagens kretsloppssamhälle (Our finest gold: Agrarian perspectives of urban technology from the mid-1800s to today's circular society). *Bebyggelsehistorisk tidskrift*, **44**, 25–38.

Mays L. W. (2010). *Ancient Water Technologies*. Springer, London.

Melkersson M. (1997). *Staten, ordningen och friheten: En studie av den styrande elitens syn på statens roll mellan stormaktstiden och 1800-talet (The state, order, and freedom: a study of ideals of state and regulation among the ruling elite in Sweden c. 1660–1860)*. Doctoral dissertation, Acta Universitatis Upsaliensis. Uppsala, Sverige.

Melosi M. V. (2000). *The Sanitary City: Urban Infrastructure in America from Colonial Times to the Present*. Johns Hopkins University Press, Baltimore, MD.

Michanek G. and Björkman U. (red.). (2003). *Miljörätten i förändring: en antologi (Environmental law in a state of change: an antalogy)*. Iustus, Uppsala.

Montelius J.-O. (1991). *Vägunderhåll och vägbygge vid 1800-talets mitt. Ett bidrag till väghållningens historia (Road maintenance and construction in the middle of the 1800s. A contribution to the history of road maintenance)*. Daedalus. Stockholm, Sverige.

Nelson M. C. and Rogers J. (1994). Gleaning up the cities: application of the first comprehensive public health law in Sweden. *Scandinavian Journal of History*, **19**(1), 17–39, https://doi.org/10.1080/03468759408579268

Newman J. and Clarke J. (2009). *Publics, Politics, and Power: Remaking the Public in Public Services*. Sage, Los Angeles, CA.

Nilsson L. (1989). *Den urbana transitionen. Tätorterna i svensk samhällsomvandling 1800–1980 (The urban transition. Urban areas in Swedish societal change)*. Stockholm.

Nilsson D. (2011). *Pipes, Progress, and Poverty: Social and Technological Change in Urban Water Provision in Kenya and Uganda 1895–2010*. PhD thesis, KTH (Royal Institute of Technology), Stockholm.

Nilsson D. (2014). *Reflections on Past Approaches and Policies for Water and Sanitation in Cities: Transformative Shifts and Future Perspectives*, Background Paper for UN Habitat Global WSS Report.

Nilsson D. and Blomkvist P. (2020). Is the "self-read water meter" a pro-poor innovation? Evidence from a low-income settlement in Nairobi. *Utilities Policy*, **68**(2021), 1–9.

Nilsson L. and Forsell H. (2013). *150 år av självstyrelse: kommuner och landsting i förändring (150 years of independant rule: municipalities and counties in transition)*. Sveriges kommuner och landsting, Stockholm.

Nogueira A., Ashton W., Teixeira C., Lyon E. and Pereira J. (2020). Infrastructuring the circular economy. *Energies*, **13**(7), 1805, https://doi.org/10.3390/en13071805

Nydahl E. and Harvard J. (2016). Den nya statens ansikten (The faces of the new state). In: *Den nya staten: Ideologi och samhällsförändring kring sekelskiftet 1900 (The new state: ideology and societal change around the turn of the century 1900)*, E. Nydahl and J. Harvard (eds), Nordic Academic Press, Lund, pp. 9–23.

Nygård H. (2004). *Bara ett ringa obehag? Avfall och renhållning i de finländska städernas profylaktiska strategier 1830–1930 (Just a slight discomfort? Waste and cleanliness in Finish cities' prophylactic strategies 1830–1930)*. Diss., Åbo Akademi, Åbo.

Olsson L. O. (2000). Technology Carriers: The Role of Engineers in the Expanding Swedish Shipbuilding System. Diss., Chalmers tekn. högsk, Göteborg.

Østby P. (1995). *Flukten fra Detroit. Bilens integration i det norske samfunnet (The escape from Detroit. The integration of the car into Norwegian society)*. Trondheim.

Ostrom E. (1990). *Governing the Commons. The Evolution of Institutions for Collective Action.* Cambridge University Press, Cambridge, UK.

Osvald T. (2022). Stadens gränsplatser: Kungliga Poliskammaren och vardagens omstridda rum i Stockholm, 1776–1835 (Urban border areas: *the royal* police chamber and everyday spaces in Stockholm). Diss., Uppsala universitet, Uppsala.

Palme S. U. (1962). *Hundra år under kommunalförfattningarna, 1862–1962: en minnesskrift (A hundred years under communal care, 1862–1962: a commemorative script).* utg, Stockholm.

Person K. M. (1999). *Några tankar om Malmös vattenförsörjning under medeltid och renässans (Reflections on Malmös water supply throughout the middle ages and the renaissance).* Malmö Fornminnesförenings årsskrift, Elbogen.

Persson S. (2008). Gerhard Oestreich, den tidigmoderna staten och det svenska forskningsläget (Gerhard Oestreich, the early modern state and the state of Swedish research). *Scandia,* **73**(1), 57–73.

Persson K. M. and Winnfors E. (2007). *Malmö – den törstande staden (Malmö – the thirsty city).* Ohlson and Winnfors, Örebro.

Petersson F. (2005). *Vattnets vägar – från vik till innergård på Södermalm 1880–1920 (Ways of water – from bay to courtyard in Södermalm 1880–1920).* Center for Health Equity Studies (Chess), Stockholm.

Pettersson O. (1988). Byråkratisering eller avbyråkratisering: administrativ och samhällsorganisatorisk strukturomvandling inom svenskt vägväsende 1885–1985 (Bureaucratization or de-bureaucratization: *administrative* and societal change in the Swedish road sector 1885–1985). Diss., Uppsala universitet, Uppsala.

Pettersson R. (ed.). (2008). *Bekvämlighetsrevolutionen: Stockholmshushållen och miljön under 150 år och i framtiden (The comfort revolution: Stockholm's households and environment during 150 years and in the future).* Stockholmia, Stockholm.

Pietilä P. E. (2005). Role of municipalities in water services in Namibia and Lithuania. *Public Works Management & Policy,* **10**(1), 53–68, https://doi.org/10.1177/1087724X05280756

Pietilä P. E. (2006). *Role of Municipalities in Water Services.* Tampere University of Technology, Tampere.

Reid D. (1991). *Paris Sewers and Sewermen. Realities and Representations.* Harvard University Press, Cambridge, MA.

Roseen P. (2020). *Rapport: Erfarenheter och analys från förvaltning av VA-samfälligheter (Rapport: experienced and analysis from the care of VA-communities),* Rapport: GEMVA 1, Vatteninfo Sverige AB, Norrtälje.

Rosen G. (2015). *A History of Public Health.* Rev., expanded ed., Johns Hopkins University Press, Baltimore, MD.

Saltzman J. (2012). *Drinking Water: A History.* Overlook Duckworth, New York.

Schalling E. (1932). Utredning angående väghållningsbesväret i städerna (Investigation regarding road maintenance challenges in cities), Bilaga till Betänkande med förslag till lag om allmänna vägar och lag vägdistrikt, m.m. Avgivet av 1929 års vägsakkunniga, Stockholm, 21.

Scheidel W. (2013). *Disease and Death. The Cambridge Companion to Ancient Rome (Cambridge Companions to the Ancient World, pp. 45–59).* Cambridge University Press, Cambridge, UK.

Schivelbusch W. (1984). *The History of Rail Travel. About the Industrialization of Space and Time During The Nineteenth Century.* Malmö.

Schön L. (2014). *En modern svensk ekonomisk historia: tillväxt och omvandling under två sekel (A modern Swedish economical history: growth and change during two centuries).* Studentlitteratur, Lund.

Schreiber H. (1961). *The History of Roads. From Amber Route to Motorway*. Barrie and Rockliff, London, UK.

Seppälä O. and Katko T. (2009). Management and organization of water and sanitation services: European experiences. In: *Water and Sanitation Services: Public Policy and Management*, E. Castro and L. Heller (eds), Earthscan, pp. 86–103, Earthscan, Oxford, England.

Sheiban H. (2002). *Den ekonomiska staden: stadsplanering i Stockholm under senare hälften av 1800-talet (The economical city: urban planning in Stockholm during the latter half of the 1800s)*. PhD thesis, Stockholm University, Stockholm.

Shiva V. (2003). *Krig om vattnet: plundring och profit (The war on water: plunder and profit)*. Ordfront, Stockholm.

Simmons D. (2006). Waste not, want not: excrement and economy in nineteenth century France. *Representations*, **96**(1), 73–98, https://doi.org/10.1525/rep.2006.96.1.73

Sjöstrand Y. S. (2014). *Stadens sopor: Tillvaratagande, förbränning och tippning i Stockholm 1900–1975 (The waste of the city: reuse, incineration and disposing in Stockholm 1900–1975)*. PhD dissertation, Lund Universitet, Lund.

Smith D. (red.). (1999). *Water-Supply and Public Health Engineering*. Ashgate Variorum, Aldershot.

Söderholm K. (2007). *Uppbyggnaden av Luleås VA-system vid sekelskiftet 1900. En djärv "miljö"-satsning i en tid av teknisk och vetenskaplig omdaning (The construction of Luleå's VA-system at the turn of the century 1900. A daring "environmental" investment in a time of technical and academical change)*, Research report, Luleå tekniska universitet, 2007:13.

Söderholm K. (2012). *When Infrastructure-Related Risk-Taking Moves from the Local to the National Level: The Planning and Construction of Centralized Water and Sewer Systems in Two Municipalities in Northern Sweden 1900–1950*, Research report, Luleå University of Technology, Luleå.

Söderholm K. (2013). Governing socio-technical transitions: historical lessons from the implementation of centralized water and sewer systems in Northern Sweden, 1900–1950. *Environmental Innovation and Societal Transitions*, **7**, 37–52, https://doi.org/10.1016/j.eist.2013.03.001

Söderholm K., Vidal B., Hedström A. and Herrmann I. (2022). Flexible and resource-recovery sanitation solutions: what hindered their implementation? A 40-year Swedish perspective. *Journal of Urban Technology*, **30**(1), 23–45.

Steffen, *et al.* (2015). Planetary boundaries: guiding human development on a changing planet. *Science*, **347**(6223), 736–746.

Stockholm Waterworks. (1961). *Stockholm Waterworks 100 Years 1861–1961*. Stockholms gas och vattenverk, Stockholm.

Summerton J. (1998). Stora tekniska system. En introduktion till forskningsfältet (Large technical systems. An introduction to the field of research). In: *Den konstruerade världen: Tekniska system i historiskt perspektiv (The constructed world: technical systems in a historical perspective)*, P. Blomkvist and A. Kaiser (eds), Brutus Östlings Bokförlag Symposium, Stockholm, pp. 19–43.

Svedinger B. (1989). *Stadens tekniska infrastruktur: en kunskapsöversikt. Statens råd för byggnadsforskning (The technical infrastructure of the city: a knowledge overview. The State Board for Construction Research)*. Stockholm.

Syssner J. and Jonsson R. (2020). Understanding long-term policy failures in shrinking municipalities: examples from water management system in Sweden. *Scandinavian Journal of Public Administration*, **24**(2), 3–19, https://doi.org/10.58235/sjpa.v24i2.8611

Takala A. J., Arvonen V., Katko T. S., Pietilä P. E. and Åkerman M. W. (2011). The evolving role of water co-operatives in Finland. *International Journal of Co-Operative Management*, **5**(2), 11–19.

Tällberg E. (2018). *Statligt och kommunalt väghållaransvar: En studie kring allmän väghållning och kommunala väghållningsområden (State and municipal road maintenance responsibility: a study on public roads and municipal street maintenance)*. Masters thesis, KTH, School of Architecture and the Built Environment (ABE), Real Estate and Construction Management, Stockholm.

Tarr J. (1999). The separate vs. combined sewer problem: a case study in urban technology and design choice. In: *Water Supply and Public Health Engineering*, D. Smith (ed.), Ashgate Variorum, cop., Aldershot, UK, pp. 289–320.

Tarr J. A. (1996). *The Search for the Ultimate Sink: Urban Pollution in Historical Perspective*, 1st edn. Akron. University of Akron Press, Akron, OH.

Tarr J. A., McCurley J., McMichael F. C. and Yosie T. (1984). Water and wastes: a retrospective assessment of wastewater technology in the United States, 1800–1932. *Technology and Culture*, **25**(2), 226–263, https://doi.org/10.2307/3104713

Tempelhoff J. (2005). *African Water Histories: Transdisciplinary Discourses*. Northwest University, Johannesburg, South Africa.

Thelle M. (2019). Stofskifte under tryk: Vandets infrastruktur og rum i København (Metabolisme under pressure: the infrastructure of water and space in Copenhagen). *TEMP - tidsskrift for historie*, **9**(18), 79–96.

Tjulin R. (2002). *I kommunalteknikens intresse: svenska kommunal-tekniska föreningens verksamhet under 100 år (In the interest of the municipal technician: The Swedish Association of Municipal Engineers)*. Masters thesis, KTH (Royal Institute of Technology), supervisor Pär Blomkvist, Stockholm.

Trottier J. and Slack P. (2004). *Managing Water Resources Past and Present*. Oxford University Press, Oxford, England.

Tullgren E. (2002). *Svenska kommunal-tekniska föreningen 100 år: [1902–2002] (The Swedish Association of Municipal Engineers 100 years)*. Fören, Stockholm.

Tvedt T. and Jakobsson E. (2010). *Introduction: Water History Is World History. Ideas of Water from Ancient Societies to the Modern World*. I. B. Tauris, London, UK.

Tvedt T. and Østigård T. (ed.). (2010). *Ideas of Water from Ancient Societies to the Modern World*. I. B. Tauris, London, UK.

Vidal B. (2022). *Small Sanitation Systems – Treatment Efficiency, Sustainability and Implementation*. PhD dissertation, Luleå University of Technology, Luleå, pp. 54–56.

Wedin R. and Björlund K. (2002). *Vatten i Stockholm, 750 år med vatten i en huvudstad (Water in Stockholm, 750 years of water in a capital city)*. Stockholms Miljöcenter, Stockholm.

Westholm G. (1995). Gaturenhållning, avfallshantering och stadsplanering. Medeltida teknik belyst av visbyfynd (Street cleaning, waste management, and urban planning. Medieval technology in light of Visby findings). *Nordisk arkitekturforskning*, **8**(1), 7–18.

Westlund H. (1992). *Kommunikationer, tillgänglighet, omvandling. En studie av samspelet mellan kommunikationsnät och näringsstruktur i Sveriges mellanstora städer 1850–1970 (Communications, availability, change. A study of the interplay between communication networks and industrial structure in Swedish medium-sized cities 1850–1970)*. Umeå.

Westlund H. (1998). *Infrastruktur i Sverige under tusen år. 1. [uppl.] (Infrastructure in Sweden during one thousand years)*. Riksantikvarieämbetet, Stockholm.

Wetterberg O. and Axelsson G. (1995). *Smutsguld & dödligt hot: Renhållning och återvinning i Göteborg 1864–1930 (Dirty gold and mortal threats: cleanliness and recycling in Gothenburg 1864–1930)*. Göteborg.

Wiell K. (2018). Bad mot lort och sjukdom: den privathygieniska utvecklingen i Sverige 1880–1949 (Bathing against dirt and disease: the private hygienic development in Sweden 1880–1949). Diss., Uppsala universitet, Uppsala.

Wikander Ö. (red.). (2000). *Handbook of Ancient Water Technology*. Brill, Leiden.

Wiking-Faria P. (2009). *Freden, friköpen och järnplogarna: drivkrafter och förändringsprocesser under den agrara revolutionen i Halland 1700–1900 (Peace, buy-outs and iron ploughs: driving forces and change processes during the Agrarian Revolution in Halland 1700–1900)*. Diss., Göteborgs universitet, Göteborg, 2010.

Wikman P. (2019). Kulturgeografin tar plats i välfärdsstaten: vetenskapliga modeller och politiska reformer under efterkrigstidens första decennier (Cultural geography in the welfare state: scientific models and political reform during the first decades of the post-war period). Diss., Uppsala universitet, Uppsala.

Winnfors E. (2008). *Sundsvall – vattenstaden (Sundsvall – the water city)*. Ohlson & Winnfors, Örebro.

Winnfors E. (2017). *Jakten på Gävles vatten. 1. Uppl (The hunt for the water of Gävle)*. Ohlson & Winnfors, Örebro.

Wittfogel K. A. (1957). *Oriental Despotism; a Comparative Study of Total Power*. Random House, New York.

Yamada M. (1981). *Roads, an Essential Aspect of Human Life. The Wheel Extended*, A Toyota Quarterly Review, Special 10th Anniversary Issue.

IWA Publishing's authorised EU representative for General Product
Safety Regulations is Diane D'Arras, 15 rue Duret, 75116 Paris,
France, e-mail: safety@iwap.co.uk.

Printed and bound by CPI Group (UK) Ltd, Croydon, CR0 4YY

12/05/2026

02108870-0010